Metalworking through History

Recent Titles in
Handicrafts through World History

Needlework through History: An Encyclopedia
Catherine Leslie

Jewelrymaking through History: An Encyclopedia
Rayner W. Hesse

Handicrafts through World History

METALWORKING THROUGH HISTORY

An Encyclopedia

Ana M. Lopez

GREENWOOD PRESS
Westport, Connecticut • London

Library of Congress Cataloging-in-Publication Data

Lopez, Ana M.
Metalworking through history : an encyclopedia / Ana M. Lopez.
p. cm. — (Handicrafts through world history, ISSN 1552–8952)
Includes bibliographical references and index.
ISBN 978–0–313–33654–6 (alk. paper)
1. Metal-work—Encyclopedias. 2. Metal-work—History. I. Title.
TS204.L67 2009
745.56—dc22 2009000192

British Library Cataloguing in Publication Data is available.

Library of Congress Catalog Card Number: 2009000192
ISBN: 978–0–313–33654–6
ISSN: 1552–8952

First published in 2009

Greenwood Press, 88 Post Road West, Westport, CT 06881
An imprint of Greenwood Publishing Group, Inc.
www.greenwood.com

Printed in the United States of America

The paper used in this book complies with the Permanent Paper Standard issued by the National Information Standards Organization (Z39.48–1984).

10 9 8 7 6 5 4 3 2 1

Contents

List of Entries

Preface

This encyclopedia focuses on the materials, processes, people, places, and cultures of metalworking. It is intended to provide a general overview of these subjects as an entrance point to those not familiar with the topic. Its purpose is to give the reader an understanding of metalworking as a result of both the inherent properties of metals as well as the societies who have worked them. The significance of these materials for the people who worked them varied widely in different cultures and movements due to such things as availability, religious beliefs, or physical properties.

There are many resources available about histories of one or two particular metals, their geological significance, or metals as an expression of technical development. This text is different in that it embraces aspects of both the overt and symbolic functions of metalwork as well as the developments in all the pertinent metals during a particular period or culture. Consideration is given to the raw materials as well as the finished products with utmost respect for the methods of transformation.

This work will be of value to nonspecialist learners with general interests and as a point of entry for further research. It will be particularly valuable to those unfamiliar with the specialized terms related to metalworking methods. The broad-ranging overview that it provides is an excellent introduction to the history of metalworking as a whole. Readers may search within it as a reference work or browse for subjects that pique their curiosity.

This book is not intended to provide directions for techniques but rather explain processes with the aim of enriching the reader's appreciation of them when encountered in the context of a culture, time, or place. For instance, one may have a greater estimation of the raised copper vessels executed in Egypt before 2000 B.C. if one understands what goes in to acquiring and technically manipulating the metal. The entries regarding processes may also be used independently as a resource for the language of techniques.

Sixty entries comprise this encyclopedia with additional information given in the form of interspersed sidebars. The entries cover major cultures with histories of metal craft manufacture, major periods and movements associated with Western metalwork, the basic techniques and materials used by metalsmiths, and a few significant practitioners representing a variety of times and regions. Entries about periods include start and end dates, whereas the entries on people have their birth and death dates or dates of known professional activity. Information about cultures has been presented in general terms as an introduction to the subject. There is a bias toward the pre-industrial production of cultures as the globalization of metalworking has created a certain amount of homogenization.

Many people who practice metalsmithing have been anonymous. This makes it difficult to highlight individuals from non-Western cultures where the emphasis on individual accomplishments in the arts may not have been as codified as it has been elsewhere. Entries about individual practitioners were chosen for many reasons, including the high level of technical virtuosity that they achieved (Nicholas of Verdun, Samuel Yellin) or their participation in the culmination of a practice (Paul de Lamerie). Others are incorporated because of their contributions as craftsmen during significant moments in history (Paul Revere, Peter Carl Fabergé). Yet others serve as examples of an archetype (June Schwarcz, John Pearson).

Entries are followed by suggested further readings, which may include books, journals, and Web sites from which to gain additional insight into the subject of the entry. Following all the entries is a bibliography that includes a selection of general works on metalworking, technical references, Web sites for related professional organizations, and the Web sites of a large selection of museums with metalwork of note in their collections.

In the time periods covered, effort has been made to cover the significant developments and at least a sampling of its practitioners from the earliest material manipulation (First Cities) to recent occurrences in handwork (Studio Movement). This extends from around 8000 B.C. through the twentieth century. The metalworking techniques that are covered in this book address the same time period, effectively including nearly every form of metal manipulation used since metals were discovered. Because the author also has a practical working knowledge of metalsmithing, these entries are written with the advantage of firsthand experience.

Objects of personal adornment, such as jewelry, are not the subject of this work. Contemporary mass production is also beyond the subject of this book, although the preliminary stages of such things, like stamping and die forming (which are also used contemporarily by artisans), have been included. In keeping with the scope and intent of this work, object types such as fabricated steel sculpture and architecture have been excluded. The general orientation of the entries takes into account that this work has been conceived as a reference for readers from Western cultures. It therefore follows the established pattern of inclusion of non-Western cultures in general and as they are appropriate to such an overview.

Entries are arranged in alphabetical order. The basic information regarding individual metals and their properties are all in one entry, "Metals," in order to give the benefit of that context. Their uses are discussed within other entries as appropriate. The interrelated nature of these topics is exemplified by the cross references in bold type throughout each subject. These will lead the reader to other entries from which tangential information may be obtained. Each entry also ends with a suggestion of related topics that may be of additional interest.

In an undertaking such as this, there are always more people involved than the author. Thanks are due to Cynthia Williams and the remarkable faculty of the Corcoran College of Art and Design and the Smithsonian Associates Program in the History of American Decorative Arts for my preparation. The faculty and staff of the College of Visual Art and Design of the University of North Texas have also been remarkably supportive. Debra Adams of the Greenwood Publishing Group has been an invaluable and patient mentor through this process. I am forever indebted to my parents, Carlos and Eleanor Lopez, for their love and support. Endless thanks and love are also due to my husband, Keith Annis; he is the best cheerleader a person could hope for.

Timeline

Before 5000 B.C. The first use of copper by man occurs between 13000 and 8000 B.C.

The chalcolithic age begins in Western Asia around 6500 to 5500 B.C.

5000–4000 B.C. Copper is being worked into sophisticated forms in the Middle East by 5000 B.C.

Metalworking from Asia is introduced along the Nile River in North Africa by 4000 B.C.

Meteoric iron is used by man as early as 4000 B.C.

Silver and gold are being made into ornaments as early as 4000 B.C.

Copper may have been smelted from ore as early as 4000 B.C.

3999–3000 B.C. Sumerians are using copper alloys, silver, and gold by 3500 B.C.

Arsenic bronze is used as early as 3500 B.C.

Rudimentary casting is being done in **Egypt** and Sumer by 3500 B.C.

Tin bronze (true bronze) appears around 3000 B.C.

Iron swords may have been made as early as 3000 B.C.

Records of lead use dates back to 3000 B.C.

Evidence of the first rivets comes from 3000 B.C.

Around 3000 B.C. people begin actively **mining** for metals.

2999–2000 B.C. Egyptians are able to produce **raised** copper vessels and gold foil that was .005 millimeters thick by the third millennium B.C.

Lost wax **casting** is developed by the third millennium B.C.

	Conscious alloying is being done in **Egypt** and Mesopotamia by 2500 B.C.
	Cupellation is being used to render silver around the Mediterranean by 2500 B.C.
	Most of the ancient world is using metals by the year 2000 B.C.
1999–1000 B.C.	Bronze has been cast in **China** since the second millennium B.C.
	Evidence of lost wax casting with hollow cores has been found in **Egypt** dating to ca. 1800 B.C.
	Egyptians learn to smelt silver from galena ca. 1600 B.C.
	Chasing and repoussé are done in metal as early as 1600 B.C.
	Gold leaf is used as early as 1500 B.C.
	The working of gold begins in Peru around 1500 B.C., the start of **pre-Columbian** metalworking in South America.
	The use of bronze is widespread in **Egypt** by 1500 B.C.
	A pewter flask from ca. 1450 B.C. has been found in **Egypt.**
	Steel is first used around 1400 B.C.
	People in **China, India,** and the Middle East acquire the ability to work iron by 1200 B.C.
	The first true **enamels,** found in mainland Greece and its islands, have been dated to between the thirteenth and eleventh centuries B.C.
	The first examples of **engraving** date to 1000 B.C.
999–900 B.C.	There is evidence of copper smelting in **Africa** from the first millennium B.C.
	The process of making brass is probably discovered after 1000 B.C.
699–600 B.C.	Lydians in West Asia Minor mint the first coins from electrum around 650 B.C.
	The oldest objects made of platinum have been dated to the seventh century B.C.
599–500 B.C.	Iron is worked in **Africa** as early as the sixth century B.C.
499–400 B.C.	Mail **armor** is being made as early as the fifth century B.C.
	The mining and smelting of iron in **China** occurs prior to the fifth century B.C.
	Ancient Greeks begin to mint coins around the fifth century B.C.
	Chinese start to use metal inlay with cast bronzes around 400 B.C.
299–200 B.C.	The Moche (200 B.C.–700 A.D.) of South America are probably the first people on the continent to develop lost wax **casting.**
0–99 A.D.	**Niello** is commonly used as an inlay for bronze, silver and gold by the **Ancient Romans** in the first century A.D., and may have been used prior to this by **Egyptians** and Myceneans.

400–499 The trans-Saharan gold trade begins in **Africa** in the fifth century.

500–599 The earliest known example of a personal eating fork is from the sixth century.

The first hallmarking system is used in the **Byzantine Empire** in the sixth century.

Persian metalwork peaks during the Sasanian dynasty, around 550.

600–699 **Enameling** is being practiced in **China** by the seventh century.

700–799 The Daibutsu (Great Buddha) of the Todai-ji Temple in Nara, **Japan,** is cast in parts in the eighth century.

900–999 **Casting** is widespread in Latin America.

1000–1099 Iron smelting is prevalent in **Africa** by the eleventh century.

1100–1199 Arab caravans bring brass to West **Africa** in the twelfth century.

King Henry II of England adopts the silver alloy of 925 parts per 1,000 from Easterling Germany in the twelfth century, beginning the sterling standard.

The use of brass in Western Europe increases in the twelfth century.

The town of Limoges in France is known for its champlevé **enamels** from the twelfth and thirteenth centuries as well as its painted enamels from the fifteenth and sixteenth centuries.

Nicholas of Verdun completes the Klosternueberg altarpiece in 1181.

1200–1299 Basse taille **enamel** is developed in Italy in the thirteenth century.

Examples of **etching** date back to the late thirteenth century.

By the thirteenth century, craft **guilds** are firmly established in Europe.

The magnificent, gothic wrought iron door hardware on the Notre Dame Cathedral in Paris is executed in the thirteenth century.

1300–1399 The first blast furnaces are developed for making iron in the fourteenth century.

Plique-à-jour **enamel** is first developed in France in the fourteenth century.

The first firearms appear in the fourteenth century in Italy and then Germany.

The **Renaissance** begins in Italy in the fourteenth century.

Zinc is being smelted in India by the fourteenth century.

En ronde bosse **enamel** (in rounded relief) is developed in the late fourteenth century.

The Worshipful Company of Goldsmiths in London receives its **guild** charter in 1327.

1400–1499	Portuguese traders bring copper and brass to **Africa** in the fifteenth century.
	Continuous drive **lathes** come into use in the fifteenth century.
	By the fifteenth century, plique-à-jour and painted **enamels** are found in Europe.
	During the fifteenth century the katana sword reaches the height of importance in **Japanese** metalworking.
	Dinant, in what is now Belgium, reaches its pinnacle as a center for brass production in the fifteenth century, just prior to being sacked in 1466 by Charles the Bold.
	Suits of plate **armor** are developed for battle by 1430.
1500–1599	The first record of the metal Paktong is from the sixteenth century.
	The Aztec empire and its riches fall to Spaniard Hernán Cortés in 1521.
	En grisaille **enameling** begins around 1535.
	While at the court of François I, between 1540 and 1543, **Benvenuto Cellini** makes his famous saltcellar.
	Italian scientist Julius Scaliger discovered platinum in 1557.
	François Briot created the Temperance Dish around 1585.
1600–1699	The seventeenth century brings the first matching sets of **flatware.**
	Bidri ware is being made in **India** by the seventeenth century.
	Pinchbeck is invented by Christopher Pinchbeck of London in the seventeenth century.
	Silversmiths are already living in **Colonial America** by 1620.
	Iron is being smelted in **Colonial America** by 1648.
	Louis XIV revokes the Edict of Nantes in 1685, resulting in many French Protestant (Huguenot) silversmiths relocating throughout Europe and America, taking the **rococo** style with them.
	Mokumé gane has been made in **Japan** from the late seventeenth century.
1700–1799	Chinese cloisonné **enamel** reaches a high point in the eighteenth century.
	From 1701 to 1708 Johann Melchior Dinglinger creates the *Court of the Grand Mogul* for August the Strong in Dresden, Saxony.
	Paul de Lamerie is commissioned to create an ewer and basin for the Goldsmith's Company of London in 1740.
	Thomas Boulsover develops the technology for making Sheffield plate in 1742.
	Nickel is isolated as an element in 1751.

	Paul Revere makes the historic Sons of Liberty Bowl in 1768.
	Titanium is discovered by Englishman William Gregor in 1791.
1800–1899	Spinning is first used in large-scale production in the nineteenth century.
	Britannia is developed as an alternative to pewter in the nineteenth century.
	The first successful electrodeposition of metal is performed ca. 1800.
	Danish scientist Hans Christian Oersted produces the first isolated aluminum in 1825.
	Tiffany & Young is established in New York in 1837 by Charles Lewis Tiffany and John B. Young. This will become **Tiffany & Company** in 1853.
	G. R. Elkington patents the process of electroplating in 1840.
	The great California gold rush begins in 1848.
	The Great Exhibition, the so-called Crystal Palace, is held in London in 1851.
	In 1853 Commodore Matthew Perry opens trade with **Japan** through a show of force, beginning a period of Japonism in the decorative arts.
	In 1859 the Comstock Lode was discovered in Nevada.
	The Bessemer steel-making process is used for the first time in the United States in 1864.
	The **Arts and Crafts Movement** is fully developed in England by 1880.
	Czar Alexander III commissions the first of many famous Easter eggs from **Peter Carl Fabergé.**
	Charles Robert Ashbee founds the Guild of Handicraft in London in 1888, with **John Pearson** as its only trained metalworker.
	Hull House, where metalwork is taught as social reform, is founded in Chicago in 1889.
	The Gorham Silver Company opens their Martelé workshop in 1891.
	Gas and arc welding are developed at the end of the nineteenth century.
1900–1999	Stainless steel is developed in Europe early in the twentieth century.
	The Paris International Exhibition of 1900 is the high point of **art nouveau.**
	The Wiener WerkStätte (Vienna Workshops) open in Austria in 1903.

Georg Jensen opens his own silversmithing shop in Copenhagen, Denmark, in 1904.

Alexander Fisher publishes his influential book, *The Art of Enameling Upon Metal* in 1906.

Samuel Yellin's workshop executes the wrought ironwork for the Federal Reserve Bank of New York from 1919 to 1924.

The **Bauhaus** school in Germany is open from 1919 to 1933.

Oxyacetylene welding is widely used by 1920.

Aluminum **anodizing** is developed for industrial use in the 1920s.

In 1926 Secretary of State Herbert Hoover limits the number of pieces in a silver set to 55.

Dr. Sōetsu Yanagi coins the term *mingei* to describe **Japanese** folk art.

The Paris International Exposition of Modern Industrial and Decorative Arts of 1925 is the high point of the **art deco** style.

The **Studio Movement** begins to coalesce after 1940.

Scandinavian **modern** design is highly influential in metalworking during the 1940s and 1950s.

At the urging of Margaret Craver, the Handy and Harmon company sponsors silversmithing workshops from 1947 to 1953.

June Schwarcz begins making her signature vessel forms after 1964.

In 1969 the Society of North American Goldsmiths, the primary professional organization of American Studio Metalsmiths, is founded.

The Artist-Blacksmiths' Association of North America is founded in 1973.

The Encyclopedia

Africa

Africa is a continent in which a variety of metals are found throughout. Most pervasive are iron, gold, and alloys of copper, all of which have been mined and manipulated with a great deal of skill. Found in smaller quantities, lead, tin, and silver have been used in limited amounts. The two methods of working metal that have been most commonly used by African peoples are **casting** and **forging.**

An overview of the main methods of working the dominant metals are given here due to the fact that the African continent is made up of many diverse peoples, all of whom have some history of metalworking. This entry will primarily focus on sub-Saharan Africa. The technologies described here generally refer to traditional methods that were widely used before the middle of the twentieth century, when Western manufacturing processes supplanted traditional ones.

IRON

Iron was being worked in portions of Africa from as early as the sixth century B.C. By around 1000 A.D., the technologies of smelting and working iron had pervaded most parts of the continent. Iron is readily available in the soil of most regions of Africa, and the methods for extracting and working it did not change much over the centuries. Surface deposits of iron ore are melted down at low temperatures over a period of two days. This happens in a furnace that is dug into the ground and topped by a walled construction above. The forms of these constructions vary by region, but they are generally domed or conical. The ore is layered with charcoal made from hardwood, and the fire is made to burn hotter through the use of bellows. When the smelting is over, a hole at the bottom of the furnace is opened, and the iron is pulled out. The result of this process is a fragile, impure iron called bloomery iron or a bloom. The bloom is then hardened and made malleable through repeated hammering and reheating before it is worked into its final form by a blacksmith. Depending on the regional traditions, the blacksmith may or may not also be the one who smelts the iron.

The seemingly magical process of extracting a useful material from apparently useless rock gives the blacksmith a unique position in African society. Frequently, blacksmiths are held in high regard as intermediaries between humans and supernatural forces. However, in some areas blacksmiths are hated and feared for these abilities and are forced to live on the edges of settlements.

In Africa, only men control the production of iron, which is likened to the process of giving birth. For this reason, there are many restrictions and taboos surrounding the smelting process that deal largely with sexuality and the drawing of blood. A man will not have sexual relations the night before he is to smelt iron. Women who are of childbearing age or who are menstruating are not permitted within the workshop. In addition to the magical nature of the smelting process itself, there is an acute awareness that steel can be used to kill, and therefore no form of violence is permitted within the workshop. Besides weapons, the processed steel is used to make a wide array of articles including religious implements, sculpture, currency, and domestic objects like oil lamps.

AFRICAN CURRENCY

Before colonization, Africans bartered for goods instead of using coins. Gold was traded in the form of dust between the peoples of West Africa and the trans-Saharan traders. As a result, the peoples of West Africa used gold dust as currency. Specially made bronze weights and spoons that were handed down from generation to generation were used by the Ashanti of Ghana to measure and store this precious commodity.

Iron was in even greater demand than gold, however, because of its strength. Iron bars became the standard measure of trade in Gambia in the nineteenth century. Along the Ubangi River in central Africa, iron was traded in the form of bells, and in Northern Zaire the trading units of iron were gardening hoes. New forms, such as *sompe* (winged bars used as currency on the Ivory Coast) were developed solely for the act of trade. When copper was used as barter it was traded in the form of cuff bracelets called *manilles.*

The making of weaponry in Africa culminated around 1900, before the influx of European firearms. Across the continent a wide variety of steel armaments were developed for combat and ritual. The Fang of western central Africa are known for their battle axes made with long, decorated wooden handles and blades shaped like a toucan's head, complete with a hole for the eye and a line for the beak. The throwing knives of central Africa resemble silhouettes of strange plants with many branches coming off of a central stalk. Throwing knives of this kind are particular to Africa. Other forms of African iron weaponry include swords, sabers, spears, knives, and daggers. The shapes of the weapons varied from tribe to tribe. Although steel was a common element in the blades of these weapons, local materials and aesthetics determined treatment of the handles and other decorative elements. For example, in the gold-rich regions of the Ivory Coast, blades were frequently decorated with gold. Across the continent, implements carried by leaders evolved into more abstract sculptural forms, becoming status symbols.

GOLD

South Africa is the largest gold-producing region in the world. In some parts of Africa, gold-bearing sand is found as a surface deposit. In the southeast, gold-included quartz was crushed, and the gold was extracted in clay crucibles. In addition to the use of surface deposits, mining is done. Since the fifth century, gold mined in West Africa was brought to Egypt, where trans-Saharan brokers traded it with Mediterranean merchants. Those who controlled the mining and trade of gold became wealthy and powerful overlords.

Little gold work made prior to the nineteenth century survives. This is mainly because of the inherent value placed on the material by Europeans and the ease of refashioning objects made of gold. There is also an indigenous history of ambiguous feelings toward this shiny metal that required ritual protection against evil spirits for those who engaged in its acquisition or use.

Cast gold was used as decoration on sword handles and on the apexes of staffs of some groups. A gold head on a ruler's throne might be used to represent a fallen enemy. On larger objects, gold foil would be pounded thin and then affixed to wooden supports. Gold is now largely used for objects of personal adornment.

COPPER

The earliest evidence of smelting technology in Africa is a copper smelt found dating from the first millennium B.C. in the southern Sahara region. As with gold in West Africa, control of copper production in southwestern Africa gave the people of the Luba kingdom access to a world of trade goods. It was used most often in the political centers that controlled its distribution.

Most of the copper in Africa is found in the central and southern regions. In Northern Africa, gold was traded in order to acquire copper as they had no local resources. The Dogon people believed that copper is associated with the sun. The Bini used it for ornaments early on, but later replaced it with imported brass. Copper is most commonly worked by hammering and texturing planar surfaces. Although pure copper was used for some articles, it was most often alloyed with other metals, especially lead, to improve its working properties (see **Metals**).

BRASS AND BRONZE

Alloys of copper were used for casting, with the specific ingredients varying widely by locality. Because they were generally used in the same manner and not all alloy constituencies are verifiable, it must be understood that in some cases the words "brass" and "bronze" are used interchangeably. Most true brass would have been imported, since there is little zinc to be found locally. Bronze casting continues today in areas of West Africa and in the outlying areas of major urban centers.

Some of the greatest achievements of African metalwork were cast in bronze through the lost wax method (see **Casting**). In this system, alternating coats of clay and charcoal are built up to form a strong wall around finely modeled wax form. Once this shell is complete, the wax is melted out and replaced with molten metal. This method has been used to create intricate surface details in cast vessels and sculptural objects that sometimes included replicas of small insects that are startlingly realistic. Igbo objects from the ninth century are thought to be the oldest lost wax cast objects in Africa.

Although lost wax casting was generally used for larger or more complex objects, smaller things could be made with direct casting methods. That is, small items like seeds or shells could be coated in the ceramic mold materials and burned out directly, without being replicated in wax. This only rendered solid castings, and was used by the Akan to make the bronze weights with which they weighed gold dust. The Akan casters made thousands of such weights, both with lost wax and direct casting. Many of the weights were in the forms of small sculptures that illustrated proverbs or aspects of daily life in exquisitely rendered miniature scale.

The Ife and Igbo peoples of the Niger River area are both known for their work in bronze. Their castings frequently mimic forms found in nature, including calabash gourds and seashells. The Ife are particularly well known for their remarkable cast brass heads. Cast bronze production was greatest during times of significant metal importation, such as when Arab caravans brought brass to West Africa in the twelfth century and Portuguese traders brought copper and brass in the fifteenth century.

Many amazing cast works in bronze were made for the court of Benin in West Africa, which flourished for four hundred years starting in the fifteenth century. The cast metal objects produced for the king and his court at that time are renowned for their technique and artistry, especially their ceremonial brass boxes. The halls of the palace were covered with bronze plates. Bronze bells were cast without clappers and played by striking the exteriors with sticks. They were made in the form of tapered cubes with an open bottom and were used for religious ceremonies and in battle to invoke ancestral help and protection.

See also Ancient Egypt; Armor and Arms; First Metalsmiths; Mining and Smelting

FURTHER READING

"Africa: The Art of a Continent," Solomon R. Guggenheim Museum, http://artnetweb.com/guggenheim/africa

African Art Museum of Maryland, http://www.africanartmuseum.org

Brincard, Marie-Thérèse, and Evelyn Fischel. *The Art of Metal in Africa.* New York: African-American Institute, 1982.

Lerer, Susan. *African Metalwork and Ivory.* Newport Beach, CA: Images of Culture, 1993.

National Museum of African Art, Smithsonian Institution. http://www.nmafa.si.edu

Schädler, Karl-Ferdinand, and Armand Duchâteau. *Earth and Ore: 2500 Years of African Art in Terra-Cotta and Metal.* München: Panterra, 1997.

Westerdijk, Peter. *African Metal Implements: Weapons, Tools, and Regalia: Collection of Frederick and Claire Mebel/Curated by Peter Westerdijk.* Greenvale, NY: Hillwood Art Gallery, 1984.

Aluminum. ***See*** Metals

Ancient Egypt
(ca. 3100–1100 B.C.)

Egypt is located on the northeastern corner of the African continent. It initially developed as settlements along the Nile River that were consolidated under a single king around 3100 B.C. Ancient Egypt was one of the earliest and longest-lasting civilizations in world history, spanning 2,000 years. The fertility of the Nile River made Egypt rich agriculturally. The barren regions outside of the river valley kept Egyptians separated from other settlements, and as a result Egypt developed as a largely conservative culture. With less exposure to outside influences, Egyptian metalwork changed very slowly.

Metalworking was introduced to the agrarian settlers along the Nile from Asia, with the influx of the Badarians at the end of the fifth millennium B.C. Although highly skilled metalworking evolved, most peasants continued to use stone tools long after metal replacements were developed. This was because metalworking in ancient Egypt was primarily used in the service of the two driving forces of the society: the king and his court and the rituals surrounding death. Egypt was an absolute monarchy ruled by a king called a pharaoh, who was believed to be a god.

PTAH OR PHTHAH

The Ancient Egyptians believed that various gods and goddesses influenced every aspect of life and death. They worshiped many deities, including Ptah, who was considered the patron of craftsmen and creation. He was depicted with a false beard, wearing mummy wrappings, gripping a staff, and frequently wearing an artisan's skullcap. It was believed that he created the whole world as well as the other gods. It was also believed that the floor of heaven and the roof of the sky were made of a metal plate made by Ptah. It is probable that his priests governed the metal craftsmen.

Egyptians believed in an afterlife that required considerable preparation for the burial of an individual. Craftsmen filled tombs with objects or representations of objects that people would need in the afterlife, including tools, servants, and personal care equipment. The elaborateness of preparation was dependent on the importance of the individual. Kings and queens had monuments like the Great Pyramids at Giza built, whereas lesser individuals had small tombs.

Production of burial articles provided an unending source of employment for craftsmen but also meant that a certain quantity of metals were being taken out of circulation constantly. Annual Nile flooding enriched the land of Egypt, and the river's constant source of water gave the inhabitants of the valley abundant, tradable resources. Due to this wealth, the Egyptians frequently imported raw materials for metalworking rather than processing what occurred in the region naturally. The king sponsored trade missions and armed expeditions to regions like Sinai and Nubia to trade for or forcefully take metals.

Craftsmen were part of Egypt's middle class. A craftsman's status depended more on who he served than how skilled he was. The three main employers of craftsmen were the royal factory, builders of the great temples, and the upper class. Craftspeople were usually brought to the trade through family lineage but some middle-class boys were apprenticed to a master craftsman outside their family. In hieroglyphics, metal workers are indicated by the inclusion of a metal vase called a *bedjty*.

In ancient Egypt, the metalsmiths generally worked many metals, not just one. Copper was probably the first metal worked in Egypt, followed by gold. Both copper and gold are native to the region. Copper vessels were being **raised** in Egypt as early as the third millennium B.C. Gold was primarily used for personal ornamentation. Although some objects were fabricated or **cast** in gold, it was primarily used to **gild** other materials like copper or wood. By the third millennium B.C., Egyptians were producing gold leaf less than .005 millimeters thick by pounding it between specially fitted stones.

BRONZE

Bronze was not widely used in Egypt until around 1500 B.C. There was no local supply of tin, so the Egyptians initially imported pre-smelted bronze. Later they imported the tin only, to combine it with locally **mined** copper. By 1800 B.C., the Egyptians were using the lost wax process to make cast objects with hollow centers. They were casting large bronze statues, architectural elements, and long, knifelike weapons. Typically the objects such as statues were cast with little detail, then incised with outlines of clothing and adornment. There are also a few examples of damascening on cast bronze (see **Inlay**). Stylized statues followed a standardized rule of proportion and formulaic presentation. The more important a person was, the larger he was depicted.

SILVER

In ancient Egypt silver was more scarce than gold prior to 1600 B.C. There was no naturally occurring silver in the region. When the technology of smelting silver from galena was developed, it gradually began to be traded to Egypt from Asia Minor. Electrum was thought to be a separate metal unto itself and was used by metalworkers as a contrasting color in decorative objects. Iron was another rarity in ancient Egypt. Egyptians were trading for iron by the first millennium B.C. but probably were not smelting it themselves until the sixth century B.C. Some lead was used for plumbing.

Much of the art of ancient Egypt was rendered in stone, a commodity available in variety and plenty. Quantities of metal statues were made as a part of religious observations and death rites. The upper classes had domestic wares made from metal, such as drinking vessels, jars, jugs, and bowls. Early vessels were made by bending strips of metal and riveting them together, then burnishing down the overlapping edges. Later, vessels were made by **raising** or sinking. Metal was also used for personal toiletry items such as combs, mirrors and razors.

See also Africa; Ancient Greece; First Metalsmiths

FURTHER READING

Dollinger, André. "Metals: Sources, Technologies, Uses," http://www.reshafim.org.il/ad/egypt/trades/metals.htm

Müller, Hans Wolfgang, and Eberhard Thiem. *Gold of the Pharaohs.* Ithaca, NY: Cornell University Press, 1999.

Petrie Museum of Egyptian Archaeology, University College London, http://www.petrie.ucl.ac.uk

Shaw, Ian, ed. *The Oxford History of Ancient Egypt,* new ed. Oxford: Oxford University Press, 2003.

Strouhal, Eugen. *Life of the Ancient Egyptians,* Deryck Viney, trans. Norman: University of Oklahoma Press, 1992.

University College London, *Digital Egypt for Universities,* http://www.digitalegypt.ucl.ac.uk/metal/index.html

Ancient Greece
(ca. 900–146 B.C.)

Ancient Greece is considered the birthplace of Western civilization. It was located on a peninsula extending into the Mediterranean Sea as well as the western coast of Asia Minor and some of the nearby islands. Greece was originally made up of small, largely independent city-states with a common language and culture. The city-state of Corinth was particularly known for its jewelry and metal goods, although each region practiced some metalworking.

Ancient Greek religion was populated by a group of deities who were believed to affect daily events, foretell the future, and grant favors when given offerings. The major gods were called Olympians and were believed to live on Mount Olympus. They were depicted with human moods and attributes but supernatural powers.

A conflicting attitude toward the manual toil of metalworkers and their highly esteemed products is evident in the Greek god Hephaestus. He was the god of fire, metal workers, and volcanoes. Although the details of his origins vary, it was said that he lived and worked on the island Lemnos. Supposedly, Hephaestus was looked down upon by the other Olympians because he was lame and engaged in manual labor. However, the other gods valued the metal objects that he made, which were of surpassing beauty and ingenuity. His cult was most active in industrial areas, especially in Athens, where he was frequently depicted alongside Athena, goddess of technology and wisdom.

Ancient Greek metalworkers operated both alone and in workshops with hundreds of craftsmen. Slaves made up a large percentage of the Greek population and worked in manufactories alongside citizens. Because many craftsmen were freedmen or slaves, a prejudice developed against artisans. However, their wares were held in high esteem.

Goods from western Asia and Egypt influenced early Greek metalwork. As time passed, Greek metalwork was characterized by increasingly naturalistic depictions of the human figure. Absolute realism was the aim of each artisan. Their standards of beauty were determined by concepts of idealized proportions, harmony, and rhythm. The Greek metalworkers excelled in hammer forming, **chasing and repoussé,** and **casting** (see **Raising**). They used a tin and lead alloy as a kind of low temperature solder (see **Connections**). The hammer-formed objects that they made included **armor,** vessels, and domestic objects.

WORTH ITS WEIGHT

Metals played a significant role in ancient commerce. In the early years of Greece, three-foot-long iron rods modeled after cooking spits were exchanged as an early form of currency. These were replaced by minted coins around the fifth century B.C. The first minted coins in history were made of electrum and produced in Lydia, located in western Asia Minor, around 650 B.C. The idea soon spread across ancient Greece, where the individual city-states each began to mint their own currency. Greek coin dies are considered some of the most beautiful and well crafted in history. Objects created in gold and silver, such as drinking vessels and small sculptures, were made to rounded weight values of the dominant currency. Because of this, these objects could be used in the place of large sums of money.

BRONZE

There were many sources of copper available to the ancient Greeks including the island of Cyprus, the name of which comes from the Greek word for copper, *kupros.* Some copper was used for votive offerings, but the copper-based alloy bronze was preferred because of its hardness and good casting ability. The tin used to make bronze had to be traded for over long distances from Asia Minor or Western Europe. Bronze was used for plowshares, tableware, coins, weapons, statues, cauldrons, and jewelry. In addition to being cast, bronze was also worked cold in sheets by raising or hammer forming over wooden substrates.

Some of the most magnificent pieces of metalwork to be made in ancient Greece were the large cast bronze statues. Bronze was cast extensively in open stone molds and with the lost wax process to produce solid or hollow objects. Castings were chased afterward to give additional details. In order to make them look as realistic as possible, statues were sometimes given glass or gemstone eyes as well as lips, teeth and fingernails of inlaid copper and silver. Bronze statuary in ancient Greece was used to depict gods, heroes, philosophers, athletes, and rulers. They stood in public areas and temples dedicated to various gods. Relatively few ancient Greek bronzes survive to this day, but the carved marble replicas that were made after the fact give us some idea of what they looked like. The influence of Greek statuary on public art continues to this day.

GOLD

Cast gold was rarely used to make objects. Instead, the Ancient Greeks would hammer the refined metal ingots into sheets. The sheets of gold were then used for death masks, wreaths, vessels, jewelry, coins, and gold leaf. Gold was not plentiful in the lands of ancient Greece, so much of it had to be traded for around the Mediterranean.

SILVER

Silver was more available locally than gold. Natural mineral deposits of silver could be found in such places as the publicly owned mines in Laurion, near Athens. It was used for jewelry, wreaths, vessels, statuary, tableware, and was the primary metal for coinage. Having silver cups and plates was considered a sign of gentility in Athens.

LEAD

Lead is a byproduct of the cupellation process that renders silver from ore. The Greeks used lead to make anchors, figurines, cheap jewelry, and weights. They also employed it as a construction material in the sheathing of ships and building joinery.

IRON

Iron was used as a substitute for bronze when tin could not be traded for. The ancient Greeks did not have the technology to cast iron, so they relied on **forging** and welding to produce tools and weapons from this material (see **Connections**). Iron was cheaper and easier to get locally than bronze, so only its inability to be cast acted as a drawback to its use. Because of the technologies available, the iron and bronze that were being rendered at the time were equally tough.

See also Ancient Egypt; Ancient Rome; First Metalsmiths; Renaissance; Rococo and Neoclassical

FURTHER READING

Ancient-Greece.Org, http://www.ancient-greece.org

Durando, Furio. *Splendours of Ancient Greece.* London: Thames and Hudson, 1997.

Garland, Robert. *The Daily Life of the Ancient Greeks.* Westport, CT: Greenwood Press, 1998.

The Metropolitan Museum of Art, New York: The Greek Galleries, http://www.metmuseum.org/explore/Greek/Greek1.htm

Oracle Education Foundation: ThinkQuest Library: Bronze, http://library.thinkquest.org/23492/data/bronze.htm

Sacks, David. *Encyclopedia of the Ancient Greek World,* Oswyn Murray, editorial consultant, revised ed. New York: Facts On File, 2005.

Treister, Michail Yu. *The Role of Metals in Ancient Greek History*. New York: E. J. Brill, 1996.

Ancient Rome
(753 B.C.–476 A.D.)

According to legend, the brothers Romulus and Remus founded the city of Rome in 753 B.C. The city grew into a massive empire and by 200 A.D. stretched west into Britain and the Iberian Peninsula, south across northern **Africa,** and east into west Asia. Under the reign of Augustus Caesar, the culture and customs of **ancient Greece** were absorbed by Rome and spread with further conquests. Trade by sea and road thrived, bringing an exchange of materials and goods as well as ideas and belief systems.

MONEY MAKERS

When Rome was first founded, payment for goods or services was made by barter. The standard by which the worth of all goods was measured was the cow. As Rome expanded and came into contact with other cultures, it became necessary to establish a more standardized form for payment. The first Roman coinage was small bits of unmarked bronze, valued for their weight. Stamped coins and a wider variety of values eventually followed. Many of the early coins copied Greek examples. In addition to bronze, currency was manufactured from brass, silver, and gold.

Coinage was minted in several locations throughout the empire, and when the imagery on a coin was changed in the city of Rome, new designs had to be sent to each outlying operation. From the fourth century A.D., the coins created in Rome were made near a temple dedicated to Juno Moneta. Juno was the Roman version of the Greek Goddess Hera, wife of Zeus and mother of the gods. "Juno Moneta" approximately translates into "Juno of the warning." People began to refer to the location of the coin manufactory as the *moneta,* which developed into the modern words money and mint.

In ancient Rome, metals were needed not only to supply the demands of Rome itself but also to fuel the campaigns of expansion. Soldiers required bronze and iron for armor and weaponry, and silver was needed to pay for the expenses of war. In addition to the copper and iron found in Italy and Greece, vast quantities of gold and silver were mined in Spain during Rome's rule. Lead and tin were mined in Britain by Roman slaves at the height of the

Empire's expansion. The metals found in the lands conquered by Rome fed its progress.

Metalsmiths traveled within the armies to make nails and arrowheads, mend hardware, and perform upkeep on the **armor and arms.** The metalworkers that traveled with the army were concerned primarily with utilitarian wares. For specialty goods or finer examples of the craft, urban centers were patronized. In addition to functional necessities, ornamental items such as parade armor, decorative sword hilts, and elaborately wrought helmets were also made. Parade armor was usually made of bronze with very fine workmanship and costly decoration, such as **gilding.** Some wrought helmets simulated hair, while others were replete with symbolism related to battle. The hardware for horses received similar treatments.

SILVER

In addition to coinage, silver was used for table wares in the affluent houses of ancient Rome. Sets of eating and drinking vessels were made for entertaining and display. Vessels and ladles for mixing wine with water, cups for drinking, serving dishes, and a few communal utensils were common at upper echelon parties. Most of the food was eaten with the hand, but special spoons were used for soups and other messy dishes. Imagery on domestic vessels might be political or mythological. Other common motifs were landscapes or the bounty of the table. The imagery on silver was modeled in the same manner as the sculpture of the day.

Important institutions such as temples and town governments would have their own collections of silver. Silver liturgical vessels were also used by the Christian church, which grew under the Roman Empire and was adopted as the state religion in 395 A.D. Silver wares were generally commissioned rather than purchased from a readymade selection. Well-made examples by significant craftsmen were highly valued and commanded high prices. However, the general degrading of the status of artisans due to the association of slavery with manual labor continued from ancient Greece into Roman times.

Silver vessels were usually created by **raising,** although **cast** elements were frequently soldered on too (see **Connections**). Decorative techniques included **chasing and repoussé,** gilding, **engraving, niello,** and the **inlay** of other metals. Interestingly, many of the early Roman vessel forms had a double wall structure. This meant that the negative impression of a repoussé exterior would be hidden by the interior shell, which was joined to the exterior element at the lip. Ideally, there would be no trace of the maker's hand in the surface of any finished piece.

If a person could not afford silver, the next most desirable materials for entertaining were glass or bronze. The bronze articles would have been formed and decorated in much the same way as silver. Indeed, even ceramic forms mimicked silver at times. In Britain under the Roman Empire, pewter was sometimes used as a substitute for silver. On rare occasions silver and gold were used for small statuary. They were also inlaid into other materials such as wood or bronze furniture.

By the second century A.D. the technical prowess displayed in earlier pieces had begun to decline. Instead of raising two silver shells, vessels were cast and the ornament merely chased. Early naturalism gave way gradually to more stylized and formalized elements. This Neoplatonist stylization valued the symbolism of forms over their realistic representation.

BRONZE

In addition to armor, bronze was used for a wide variety of useful things as well as statuary. Bronze statuary for public display was cast with the lost wax process. Sometimes other metals, such as copper and silver, were inlaid to heighten the realism. Wealthy Romans frequently adorned their furniture with metal elements. Polished, gilded, and inlaid bronze ornaments were occasionally used in this manner.

Utilitarian wares such as bronze cooking vessels might be cast in reusable stone molds. It is believed that the stone molds may have been made with a **lathe.** Such forms could also have been finished on a lathe. Occasionally, bronze vessels were made by raising sheet metal. Cooking vessels were often coated with tin on the interior to prevent food from reacting with the bronze.

BRASS AND PEWTER

The use of brass in ancient Rome increased toward the first century A.D. Pewter was used as both a low-temperature solder and for household utensils. These were cast in much the same way as bronze, usually into reusable molds. They too would have been cleaned on a lathe to make them more regular.

IRON

Although the Romans made more use of iron than the Greeks did, they were still only able to forge the metal, not cast it. Also, there remained sporadic problems with controlling the carbon content. The Romans both made and imported cast damascene steel (see **Lamination Techniques**). In addition to tools and some armor, blacksmiths made locks, cart hardware, and tools.

See also Byzantine Empire; Celtic People; Middle Ages

FURTHER READING

Boardman, John, Jasper Griffin, and Oswyn Murray. *The Oxford History of the Classical World.* Oxford: Oxford University Press, 1986.

The British Broadcasting Company, "Ancient Romans," http://www.bbc.co.uk/history/ancient/romans

The British Museum, London, "Ancient Rome," http://www.britishmuseum.org/explore/world_cultures/europe/ancient_rome.aspx

Henig, Martin. *A Handbook of Roman Art: A Comprehensive Survey of All the Arts of the Roman World.* Ithaca, NY: Cornell University Press, 1983.

Kent, John, and K. S. Painter. *Wealth of the Roman World: AD 300–700.* London: Published for the Trustees of the British Museum by British Museum Publications, 1977.

The Metropolitan Museum of Art, New York, "Greek and Roman Art," http://www.metmuseum.org/Works_of_Art/greek_and_roman_art

The Public Broadcasting Service, "The Roman Empire in the First Century," http://www.pbs.org/empires/romans

Scranton, R., J. E. Packer, D. K. Hill, and C. Roebuck. *The Muses at Work. Arts, Crafts, and Professions in Ancient Greece and Rome.* Ed. by C. Roebuck. Cambridge, MA: The MIT Press, 1969.

Strong, Donald Emrys, and David Brown. *Roman Crafts.* New York: New York University Press, 1976.

Annealing

Annealing is the process of using heat to relieve stresses within a metal. This is done in order to make a metal more malleable. The more tightly packed and regular a metal's crystalline structure is, the more rigid the material. Annealing creates irregularities in the crystalline structure. For most nonferrous metals (see **Metals**), this is done by heating the metal to approximately two-thirds of its melting temperature and quenching it (plunging it into water or another liquid). The colors that a metal turns as it is being heated may act as an indication of its temperature.

Most ferrous metals are annealed by heating to a prescribed temperature and then cooling them very slowly. The cooling process is prolonged by placing the metal in a material or container that prevents heat from escaping quickly, such as a box of lime or a controlled furnace.

Metals become hard as a result of many processes. Work hardening is the hardening of metal through one of many possible forming process such as bending, **chasing,** spinning, **forging,** or **raising.** Controlled heat may also be used to harden a metal in a process known as heat hardening. When nonferrous metals are heated to their annealing temperatures for long periods, a more perfect alignment of the crystalline structure occurs, and they loose their structural irregularities.

See also Lathe Work

FURTHER READING

McCreight, Tim. *The Complete Metalsmith,* revised edition. Worcester, MA: Davis Publications, 1991.

Anodizing

When certain metals come into contact with oxygen they form a metallic oxide. For example, when iron is exposed to oxygen it forms ferric oxide, which is also known as rust. By using certain metals as the positively charged electrode or anode when conducting electricity through a dissolved (usually acidic) substance, it is possible to build up thicker layers of oxide on the surface of those metals than would occur with normal exposure to oxygen. The process of using the metal as an anode to build up a layer of metal oxide is known as anodizing. Aluminum and titanium alloys are most commonly anodized, but the process can also be used with niobium, magnesium, and tungsten, as well as other tough grey metals and their alloys. In metals such as aluminum and titanium, that oxide layer can be used to protect and even color the surface of the metal. This process is generally used to color metals that do not usually take **patinas.**

ANODIZING ALUMINUM

The process of anodizing aluminum was developed in the 1920s for use by industry. Aluminum oxide is the same color as aluminum and takes the shape of tiny tubes standing perpendicular to the surface of the metal. Also known as alumina, it is corrosion resistant and porous. In this state it is capable of accepting dyes that are then sealed into place with hot water or steam. An anodized surface is a hard and brittle skin with a higher melting temperature than its aluminum core. Bending or striking anodized aluminum can easily mar an anodized and dyed surface. Therefore, anodized aluminum must be connected to other metals with cold joining processes or anodized after forming has taken place. Aluminum alloys form oxides of different colors than pure aluminum.

ANODIZING TITANIUM

Titanium may be colored through its oxide formation without the use of dyes. Titanium oxides of different thicknesses will result in the appearance of different colors. This is due to the eye's perception of the light bouncing around in the oxide layer. The thickness of the oxide (and therefore the color effect produced) may be built up deliberately by heat or electricity. In the heating process, one is using the higher temperature to bring about an increased reaction of titanium with oxygen. Electrically anodizing titanium can be controlled more easily, as oxide layers of varying thicknesses are produced through the use of higher or lower voltages. The colors that may be achieved comprise a large prismatic range.

See also Connections; Metals; Studio Movement

FURTHER READING

International Aluminum Institute, http://www.world-aluminium.org

Donachie, Matthew J. *Titanium: A Technical Guide,* second edition. Materials Park, OH: ASM International, 2000.

Wernick, S., R. Pinner, and P. G. Sheasby. *The Surface Treatment and Finishing of Aluminum and Its Alloys,* fifth edition. Metals Park, OH: ASM International, 1987.

Armor and Arms

Armor is protective equipment generated for defensive use in battle and arms are the weapons used offensively in similar situations. The two have developed in close relation to one another, and both have been the products of various fields of metalworking throughout history. The following is an overview of the development of metallic armor and arms in the West.

ARMOR

Metal workers who produce armor are generally known as armorers. Bronze was the first metal used to make armor in most regions. It was used to reinforce shields, helmets, and body padding made of such materials as leather and wood. The **ancient Greeks** and **ancient Romans** wore bronze or iron armor in the forms of helmets, breastplates, and shin covers.

From the fifth century B.C. through the fourteenth century A.D., the primary form of protection in battle was mail. This was made up of thousands of small loops of iron that were riveted closed in interlocking patterns. A shirt of mail would have been worn with an iron helmet and accompanied by a shield. By the eleventh century A.D. the mail shirt was joined by a mail hood, leg coverings, and mail mitten gauntlets, which were like gloves. As helmet forms developed, they began to cover more of the face, making the wearer unrecognizable. This led to the development of a system of graphic symbols known as heraldry that communicated the wearer's identity.

The introduction of the crossbow, which projected arrows with enough force to pierce mail, required a change in defensive strategy. Large plates of iron or steel were incorporated into the surface of armor in the thirteenth century. By the first third of the fifteenth century, this had developed into full suits of plate armor that were articulated at the joints of the wearer in order to facilitate movement. The protection of the plate armor was such that a shield was usually unnecessary. Plate armor was also made for the horses that were taken into battle.

In the early fifteenth century, Italian armorers began producing elaborate pieces that were supposed to recreate the battle garb of the heroes of antiquity, called *armor all'antica.* This was largely used for court ceremony. Armor making peaked in the fifteenth and sixteenth centuries, with highly regarded versions coming from both Italy and the Germanic regions.

Lavishly decorated armor was produced during the **Renaissance** for sporting and ceremonial purposes. The sporting event that necessitated the wearing of armor was the tournament. A tournament was made up of a series of combat-related exercises and was originally a way for fighters to prepare for battle. Over time it became a popular form of entertainment, and by the fourteenth century it was a reason in itself to acquire armor.

The necessity of suits of armor was diminished by the introduction of firearms. By the seventeenth century, only the helmet and breastplate remained as armor use fell into decline.

Armorers were first valued for their technical ability, as life and death frequently depended on the quality of their wares. As armor became more refined, additional value was placed on examples that included artistic decoration. In the first half of the sixteenth century, heads of state gave such artful armor to one another as gifts. At the height of the nobility's interest in armor it was not uncommon for court artists to design suits of armor or their embellishments. Hans Holbein the Younger and Albrecht Dürer both served in this capacity. Some court armorers were held in as high regard as any court artists.

As armor became more ceremonial and decorative, the ornamental techniques lavished on it grew. One of the most common forms of decoration was **etching.** The act of recording an etched design on armor by transferring it with ink to paper may have originated the use of etching for printmaking. Bluing was a method of heat **patination** that was frequently paired with **gilding** on **armor.** Additional techniques of decoration included **chasing and repoussé,** painting, lacquering, **engraving, inlay,** and the application of textiles to the surface.

ARMS

The development of bronze **casting** led to the first metal weapons, which were usually knife or axe forms. They may have played dual roles as functional household objects and weapons. As time went by, more specialized weapon types developed.

Swords

Swords consist of a long blade and a handle or hilt. They also frequently incorporate some kind of guard to protect the hand. The sword was the dominant weapon for most of European history and the preferred weapon of medieval aristocracy.

The earliest swords were made of bronze. Iron swords may have been made as early as the third millennium B.C. These were still relatively soft until steelmaking technologies improved around the middle of the second millennium B.C. These first swords would have been only one and a half to two feet long.

In the early Middle Ages, swords had straight, broad blades, a single cutting edge, and a rounded end. These were meant for cutting strokes rather than jabbing. By the eleventh century, double-edged swords had come into use. Around the start of thirteenth century, the use of the sword changed from a swinging movement to more a versatile one. A raised spine down the center of the blade provided additional strength, and the end was tapered to a lethal point. In the fourteenth century, hand-and-a-half swords and two-handed swords were also seen. Hand-and-a-half swords were gripped in one hand while the other hand manipulated the end or pommel for increased guidance.

By the end of the sixteenth century, the rapier had begun to usurp the medieval sword form. Rapiers were based on swords but were lighter, shorter, and more graceful. The use of rapiers developed with the art of fencing, which began in the fifteenth century and was popular in the sixteenth and seventeenth centuries. Rapiers were also used in gentlemen's duels in England and France. With the advent of firearms, sword use declined, although cavalry members carried swords until World War I.

Many exquisite display and presentation swords have been made with decorative hilts, blades, and hand guards. Iron chiseling was a metalworking technique used by both blacksmiths and locksmiths to form hot, plastic steel. Additional decorations within the blacksmith's area of expertise included etching and engraving. For more elaborate jewel, gold, or **enamel**-encrusted swords, the skills of the goldsmith were needed. In some localities, the work of sword making was divided up into areas of specialization by **guilds** in order to both employ more people and increase the quality of the finished product.

Daggers

Daggers are essentially a cross between knives and swords that were worn at the side of the body. They were common weapons from prehistoric times through ancient Rome. It was traditional for the Roman emperor to wear a dagger around his neck as a symbol of his power over life and death. During the Middle Ages daggers largely fell out of use, but they returned in the thirteenth century. In the fifteenth and sixteenth centuries they were a part of fashionable dress, and carrying an artfully decorated one was a status symbol.

Staff Weapons

Staff weapons are generally made of metal fittings attached to wooden poles. They include lances, spears, pikes, and halberds. As weapons they have historically been second to the sword in use. In the thirteenth century their importance grew through their employment by troops in formation. After the fifteenth century they were used more for ceremonial display by guardsmen or for hunting. A form of halberd is still used by the Swiss Guard at the Vatican in Rome.

Firearms

Firearms are weapons generally consisting of a metal tube from which the force of a controlled explosion causes the rapid expulsion of a projectile. Common firearms include guns and canons. The first firearms appeared in Europe in the middle of the fourteenth century, initially in Italy and then in Germany. Early examples were composed of tubes with one closed end and a hole for igniting the gunpowder, cast in bronze and mounted to wooden poles. The tubing was later developed in iron. Early guns required the services of a blacksmith for the barrel and a wood carver for the stock or wooden grip. By the sixteenth century a weapon might be passed through the hands of at least four specialized craftsmen including a barrel maker, locksmith, stockmaker, and gunsmith.

Methods of protecting the gunpowder and lighting it more efficiently continued to develop. By the early sixteenth century a form of handheld firearm had evolved that was relatively lightweight and easy to use. It was also at that time that surface embellishments were added with some regularity. As with swords, the firearms of the upper class were frequently bedecked and bejeweled in all aspects and might require the talents of specialty furniture makers and goldsmiths as well.

See also Africa; Byzantine Empire; Colonial America; First Metalsmiths; Flatware; Lamination Techniques; Middle Ages

FURTHER READING

Armory of the State of Styria, Austria, http://www.zeughaus.at

Arms and Armor at the Metropolitan Museum of Art, New York, http://www.metmuseum.org/Works_of_Art/arms_and_armor

Bull, Stephen. *An Historical Guide to Arms and Armor.* New York: Facts on File, 1991.

Ffoulkes, Charles John. *The Armourer and His Craft from the XIth to the XVIth Century.* New York: Benjamin Blom, 1967.

Fliegel, Stephen N. *Arms and Armor: The Cleveland Museum of Art.* New York: H. N. Abrams, 1998.

Guide to the Collections: Arms and Armor. New York: The Metropolitan Museum of Art, 1962.

Kershaw, Andrew. *Guide to Arms and Armor*. Windermere, FL: R. Rourke Publishing, 1981.

Kunsthistorisches Museum Vienna, "Arms and Armor," http://www.khm.at/system2E.html?/staticE/page145.html

Norman, A. V. B., and G. M. Wilson. *Treasures from the Tower of London: An exhibition of Arms and Armour.* Norwich, England: Sainsbury Centre for the Visual Arts, 1982.

Philadelphia Museum of Art Arms and Armor Collections, http://www.philamuseum.org/collections/216–427–334.html

Royal Armories, Great Britain, http://www.royalarmouries.org

Art Deco
(ca. 1912–1939)

Art deco, also known as *art moderne,* was an international stylistic period in the decorative arts that largely took place between World Wars I and II. The term "art deco" itself was applied to the period after the fact in the 1960s and derives from the 1925 Paris Exposition Internationale des Arts Decoratifs et Industries Modernes (Paris International Exposition of Modern Industrial and Decorative Arts). What is now known as the art deco style coalesced and peaked at the exposition.

At the time, the style was simply called *moderne* (not to be confused with **modernism**). Practitioners of art deco were making a conscious effort to create objects that were of their own time and not an amalgam of historical styles. The style radiated out from Paris, which held the world market share of luxury goods at the time. Emphasis was on custom pieces for high-end clientele, but the art deco aesthetic eventually diffused into the most basic and mass-produced household goods.

The interwar years were characterized by great strides in technology, dramatic economic changes, and international political unrest. In the decorative arts, mass production was widespread, and the rise of the professional designer as an occupation was growing the desires of consumerism. Ships, trains, airplanes, and the automobile were rapidly expanding the possibilities of travel, not just for need but also for leisure. The increasing presence of skyscrapers that began at the end of the nineteenth century was continuing to push the vertical limits of the urban horizon. The echo of World War I combined with the Roaring Twenties and events like the Great Depression made for a general feeling of unprecedented change.

Stylistically, art deco exhibited such diverse influences as new technologies, modern art, exotic cultures, and Sergey Diaghilev's Ballets Russes. The popularization of travel brought the aesthetic of streamlining, a low curving profile that reduces wind resistance on things like trains and planes. This was applied even to objects for which the ability to travel through air quickly was of no consequence, such as a teapot. High-end art deco metalwork utilized exotic and new materials such as ivory and rare woods alongside silver, gold, and platinum. In the middle-class markets, new materials such as aluminum, stainless steel, chrome plating, and plastics brought the mechanical sheen of art deco within reach of the average consumer.

The idealization of luxury also found expression through references to exotic cultures. Egyptian motifs became particularly popular after the Tutankhamen Exhibition held in Paris in 1922. **Africa,** ancient Mexico, **Japan,** and the American

NOT NEW ENOUGH

The 1925 Paris International Exposition of Modern Industrial and Decorative Arts was one of the many expositions since the Crystal Palace of 1851 (see **Arts and Crafts Movement**) to act as an international showcase for the fine and decorative arts. The organizers of the 1925 exhibition invited over 20 countries to participate. This exposition was unique, however, in that it had a requirement for participation: all works submitted had to be in the new modern style; no historicism would be permitted. Despite the economic boom of the Roaring Twenties, when Commerce Secretary Herbert Hoover consulted leading authorities, he was told that America could not participate because there was nothing appropriately modern to send. Although America had to decline the invitation to participate, a number of collectors, scholars, and curators from the United States did attend the exposition, and the pieces that they purchased and brought back with them helped to develop the modern aesthetic in America.

Indian were also looked to for decorative motifs. Furthermore, art deco showed evidence of the reduced geometries found in abstract art movements such as cubism and futurism and in the products of the **Bauhaus.** Additional motifs included highly stylized and abstracted figures, plants, animals, and the sun's rays. The results of these influences were objects of bolder proportions, austere surfaces, and rich materials designed to convey elegance and newness.

Although textural embellishment was generally eschewed in favor of highly polished surfaces, art deco metalwork also employed the Far Eastern technique of lacquering. Lacquer from this tradition is derived from tree sap that is thinned, purified, and sometimes colored before being applied. Dozens and sometimes hundreds of layers of lacquer are painted over wood or metal and dried under high heat and humidity, making it a slow and painstaking process. Seizo Sugawara, a Japanese lacquer craftsman in France, taught a number of Westerners the technique including Jean Dunand. Dunand is best known for his **raised** and lacquered metal vessels in the art deco style. His vases were finished with stylized geometric motifs executed in lacquer. Because lacquer is most easily dyed certain colors, his palette was limited to red, black, and clear lacquer. Greater variety was achieved through the incorporation of silver and gold metallic foils and *coquille d'oeuf,* embedded white eggshell fragments.

In silver, the master of art deco was Jean Puiforcat. Jean learned metalworking under the direction of his father, Louis Victor Puiforcat. At the family workshop, Jean studied the historical French forms that his father kept on hand for research and replication. When Jean left to start his own business in 1920, he began to make reductive versions of traditional forms that eventually embodied the restrained geometries of art deco. Working in silver with sparingly placed semiprecious stones and wood, his table wares were the epitome of austere luxury. Their flawless surfaces belied the painstaking traditional processes used to achieve them.

Other metalsmiths of the period embraced not only the aesthetic of the new technologies but their working capabilities as well. Edgar Brandt is an example of a blacksmith whose work blended traditional manufacture with the new tools of industry. Brandt was a major proponent of the oxyacetylene torch, which was invented in 1901 and produced temperatures high enough to weld steel without a forge. Many traditional blacksmiths eschewed this new tool, but Brandt's career was built on the successful reconciliation of opposites such as art and industry, jewelrymaking and architectural ironwork, one of a kind luxury goods with standardized catalogue pieces, abstraction and representational imagery. His workshop utilized assembly line production techniques as well as power hammers, **stamping** machines and spinning **lathes.** His legacy is a proliferation of wrought iron that was truly a testament to its time.

Although World War II brought about the end of the period of art deco, it had a lasting international impact on design in the decorative arts. Indeed, it was arguably the last of the international design movements.

See also Art Nouveau; Arts and Crafts Movement; Connections; Japan; Jensen, George; Nineteenth Century; Tiffany & Company; Yellin, Samuel

FURTHER READING

The Art Deco Society of Los Angeles, http://adsla.org

The Art Deco Society of New York, http://www.artdeco.org

Benton, Charlotte, Tim Benton, and Ghislaine Wood, eds. *Art Deco 1910–1939.* London: V&A Publications, 2003.

Delorenzo, Anthony. *Jean Dunand.* New York: E. P. Dutton, 1985.

Kahr, Joan. *Edgar Brandt Master of Art Deco Ironwork.* New York: Abrams, 1999.

Krekel-Aalberse, Annelies. *Silver 1880–1940: Art Nouveau, At Deco = Zilver = Silber.* Stuttgart, Germany: Arnoldsche, 2001.

Victoria and Albert Museum, London, "Art Deco Study Guide," http://www.vam.ac.uk/nal/guides/art_deco/index.html

Art Nouveau
(ca. 1890–1910)

Art nouveau literally translates from French as "new art." The term came to be used to describe the decorative arts objects being produced by the avant-garde in

Europe and America at the end of the nineteenth century. It was solidified as the name of the movement in France in 1895 when Sigfried Bing opened his shop in Paris, L'Art Nouveau, which showcased work in the new style.

The end of the nineteenth century was a time of great flux in society, enhanced by the increased transformation of life through new technologies such as the telephone and automobile. Designers and craftspeople were looking for a way to reflect that cultural revolution in the work they created. There was a self-conscious break from historicism across all forms of visual culture. The philosophy of art nouveau also sought equal treatment of all the arts, leading many fine artists to design work for the decorative arts.

Preceding styles that influenced the development of art nouveau included the **Arts and Crafts Movement,** which stressed the value of handwork, and the Aesthetic Movement, which promoted the making of art for its own sake. Additional influences included the Symbolist Movement in painting and writing, Sigmund Freud's publications about dream imagery and the unconscious, stylized natural motifs, and re-imagined regional traditions. The arts of **Japan, China,** and various colonies in **Africa** and Indonesia were also looked to for inspiration.

Art nouveau showed up in many countries at once and was named and interpreted differently in each place. In addition to France, early centers of art nouveau included Germany and Belgium, most notably Brussels. In Germany the new style was referred to as *Jugendstil* which means "Young Style." This nomenclature referenced a publication called *Die Jugend,* which popularized the movement in Germany. In Austria the movement was called secession style because its founders were associated with the secessionist fine art movement. In Italy the new art was called *La Stil Liberty,* after the Liberty & Co. shop in London that sold work in the new manner. In both England and America the Arts and Crafts Movement and art nouveau intermingle. The different national styles were broadcast through periodicals and a series of international exhibitions, most notably the Paris International Exhibition of 1900.

It has been suggested that the first evidences of art nouveau in metalwork were the attenuated handles designed by C. R. Ashbee for the metalwork produced by the Guild of Handicraft. Another designer whose works span both the Arts and Crafts Movement and art nouveau in England was Archibald Knox.

Although art nouveau disdained historical styles, it was greatly inspired by national traditions. For example, in France where the **rococo** style had made a major impact in the eighteenth century, it is possible to see similar asymmetrical, scrolled forms played out in the whiplash curve exhibited in the art nouveau of that country. The French led the world in the production of high-quality luxury goods at the end of the nineteenth century. Parisian silversmiths of the new style incorporated semiprecious stones and enamels into their creations. They also frequently created silver mountings for glass and ceramics wares.

GESAMPTKUNSTWERK: THE TOTAL WORK OF ART

An idea associated with the art nouveau movement is the achievement of a *gesamptkunstwerk*—or a total work of art. Architects, designers, and fine artists designed work to be executed in a range of media for use in completely orchestrated interiors. The aim was an environment where every element, even the flatware, worked in total harmony. The idea was derived from the operas of Wagner. Some of the most advanced metalwork being made at the turn of the nineteenth century in Britain, Germany, and Austria was meant to be part of a holistically designed room or building. This was also in keeping with the art nouveau aim of placing the fine and decorative arts on equal footing. Some of the most successful examples of the *gesamptkunstwerk* were created in Brussels, including Henry van de Velde's Bloemenwerf House, the Palais Stocklet by Josef Hoffman, and Victor Horta's Tassell House.

The Paris subway stations that were erected for the International Exposition were designed by Hector Guimard and have served as icons of the art nouveau style in France. Made of modular **cast** iron elements and glass, the forms that Guimard designed could be combined to produce five different kinds of station covers. The thin metal elements stretched into space giving the appearance of living wings with weblike membranes of glass stretched across them.

Cast metalwork allowed for the organic forms in French and Belgian art nouveau and was used extensively to create decorative and domestic objects. The plasticity that was made possible by the casting process allowed for the kinds of undulating surfaces used in much French and Belgian art nouveau. Forms frequently incorporated the writhing bodies of women with flowing hair and gowns, known as femme fatal figures. Belgian metalwork of the period was also characterized by vivacious linear tendrils that defied gravity and activated the space around them. Important silver designers in Brussels included Fernand Dubois and Philippe Wolfers.

In Germany, the members of the secessionist art movement opened a crafts workshop in 1898. Called the United Workshops for Art in Handicraft, it provided an outlet for artists like Richard Riemerschmid to design household goods in many media that were then executed by trained craftsmen. The output was generally characterized by simplified and stylized natural forms with subtle undulations. Fritz von Müller and his assistant Ernest Riegel were significant silversmiths in Germany at the time. In mass production, the German firm of Württembergische Metallwarenfabrik (WMF) produced affordable electroplated versions of Art Nouveau tableware (see **Electroforming**).

Inspired by C. R. Ashbee and the Guild of Handicraft, Josef Hoffmann and Koloman Moser founded the Wiener Werkstätte (Vienna Workshops) in Austria in 1903. Backed by the wealthy industrialist, Fritz Wärndorfer, artists designed objects to be handcrafted in many media. Hoffmann produced numerous designs for metalwork including baskets, vases, **flatware,** and tea sets. The objects produced by the Vienna Workshops were of pared down geometries and emphasized grid structures. Hammer marks were left on the surfaces of metal to indicate the method of hand production.

In Scotland, new forms were developing around the Glasgow School of Art. Two sisters, Frances and Margaret MacDonald, collaborated first with each other and then with their husbands, Herbert McNair and Charles Rennie Mackintosh, respectively. Together they were known as The Four. Although the sisters worked in many media, some of the most outstanding work they did were repoussé panels, frames, and plaques that they themselves executed in tin. Their work was characterized by ghostly figures and abstracted floral motifs.

In the United States at the end of the nineteenth century the Gorham Silver Company was the leading producer of silver tablewares. Although their success was partially due to the early use of mechanized processes, in 1891 a small workshop within Gorham was dedicated to the production of handmade art nouveau silver. William C. Codman, an English silversmith, directed the production. The effort required a great deal of retraining of staff and resulted in a line of silver called Martelé, which is French for "hammered."

Metalwork in the art nouveau style embodied both the newness of the age as well as various regional tendencies, from the nervous whiplash line to stark geometry. In precious metals the emphasis on the handmade meant that the new style was available only to the wealthy. Pewter, copper, and brass found some favor in tableware designs. As the twentieth century neared, ironwork was being used by avant-garde architects as a way to integrate structural and decorative elements.

Art nouveau began to loose influence in the years preceding World War I. By the second decade of the twentieth century, designers increasingly turned to issues of functionalism. Furthermore, international strife brought about a renewed interest in more traditional historical styles.

See also Art Deco; Bauhaus; Celtic People; Jensen, Georg; Modernism; Nineteenth Century

FURTHER READING

The Artists, "Art Nouveau," http://the-artists.org/movement/Art_Nouveau.html
Arwas, Victor. *Art Nouveau: The French Aesthetic.* London: Andreas Papadakis, 2002.
Bröhan Museum, Berlin, Germany, http://www.broehan-museum.de
Escritt, Stephen. *Art Nouveau.* London: Phaidon, 2000.
Fahr-Becker, Gabriele. *Art Nouveau.* Cologne: Könemann, 1997.
Greenhalgh, Paul. *The Essence of Art Nouveau.* New York: Harry N. Abrams, 2000.
Greenhalgh, Paul, ed. *Art Nouveau 1890–1914.* New York: Harry N. Abrams, 2000.
Kerssenbrock-Krosigk, Dedo von. *Modern Art of Metalwork.* Berlin, Germany: The Bröhan Museum, 2001.

Kunsthistorisches Museum, Vienna, Austria, http://www.khm.at/homeE/homeE.html
National Gallery of Art, http://www.nga.gov/feature/nouveau/exhibit_intro.shtm
Schweiger, Werner J. *Wiener Werkstatte: Design in Vienna, 1903–1932.* New York: Abbeville Press, 1984.
Weisberg, Gabriel P. *Art Nouveau Bing: Paris Style 1900.* Exhibition catalogue. New York: Abrams, 1986.
Weisberg, Gabriel P., Edwin Becker, and Évelyne Possémé, eds. *The Origins of L'Art Nouveau: The Bing Empire.* Exhibition catalogue. Amsterdam: Van Gogh Museum, 2004.
Whitehouse, Richard, "Art Nouveau to Art Deco," http://www.modernsilver.com/artnouveau deco.htm

Arts and Crafts Movement
(ca. 1880–1916)

The Arts and Crafts Movement originated in Britain in the mid-nineteenth century and has had a significant impact on the decorative arts. The ideals of this movement developed in response to the rapid rise of industrialization and the perceived aesthetic and social ills that mass production brought with it. Arts and crafts practitioners wished to reform the standards of design and production of goods, transforming labor into something pleasurable and rewarding that in turn, they believed, would produce goods that were simple and useful and therefore beautiful.

Britain was the first country to undertake industrialization on a large scale; thus it was there that the arts and crafts response to it began. Fully developed by 1880, the movement soon spread to other countries, most notably America and Germany. The name of the movement came from the Arts and Crafts Exhibition Society, which was started in Britain in 1887 in order to provide a regular exhibition venue for the applied arts. In each country the Arts and Crafts Movement influenced, it was adapted to incorporate the regional heritage and resources of that region.

Nineteenth century industrial manufacturing processes allowed for large quantities of goods to be made cheaply. Decorative techniques that once had taken a skilled laborer many hours to form with hand tools were now **cast** or die formed in a matter of minutes by unskilled laborers running machines for low wages (see **Stamping and Die Forming**). Objects soon became covered with a profusion of decorative motifs taken from all manner of historical styles. The replacement of handwork by machines, excess of unnecessary ornamentation, and poor working conditions of the factory were what the Arts and Crafts Movement was trying to change.

Arts and crafts practitioners valued simplicity of form and decoration. They believed that makers should be true to their materials and not attempt to make them appear to be that which they are not. It was believed that morally sound, pleasurable work would inherently produce beautiful objects and that owning such objects would improve the lives of those who used them. To this end they saw the decorative arts as being on equal footing with the fine arts.

Although the Arts and Crafts Movement was more of a philosophy of making than a style, its dictating principles did often result in similar manifestations by different makers. Arts and crafts objects were designed with their function in mind. They had simple profiles that frequently drew inspiration from nature and regional folk art. The use of less precious materials was more in keeping with their ideals. Although some arts and crafts silver work was done, a great deal of the metalwork was executed in copper and other less expensive metals. The handwork that went into shaping them was believed to make them precious.

THE DESIGN REFORM MOVEMENT

The Design Reform Movement, like the Arts and Crafts Movement, was a reaction in opposition to the early products of mass production that began in Britain. Rather than prescribe a return to handwork as arts and crafts had done, the design reformers sought to improve the standards of design for industry. Their aim was to educate designers and the growing middle class about taste and style in order to direct them away from the exuberance and eclecticism of the nineteenth century. One of their greatest triumphs was the first International Exhibition held in 1851, also known as the Crystal Palace Exhibition, at which were displayed manufactured goods from around the globe.

Members of the reform movement included A. W. N. Pugin, Henry Cole, Richard Redgrave, Prince Albert, and Owen Jones. These men were mentors to Christopher Dresser, who had attended a newly established Government School of Design. Dresser's designs embodied all that the reformers had hoped to achieve: he rejected the revival styles and studied nature, taking from it a sense of geometry and abstraction. His designs for geometric, electroplated metalwork executed in the nineteenth century look futuristic even to this day.

A roughly planished surface was frequently used on arts and crafts movement metalwork (see **Raising**). In some cases, planishing was so blatantly applied as to be counter to the true use of planishing, which is to smooth a surface. It was a mark of the craftsman, and therefore its presence was a signifier of handwork. Materials used for metalwork included copper, brass, silver, bronze, iron, and steel.

Interest in **enameling** was also revived during the Arts and Crafts Movement for it was a more affordable means of adding color than with gemstones. Also, enamel could be handled in a painterly manner, potentially achieving the arts and crafts ideal of unifying the fine and decorative arts. Arts and crafts groups made instruction in metalwork and enameling available to a wide section of society and helped to ensure that handwork techniques for metal would not be lost to history.

One of the founders and most outspoken advocates of the ideals of the Arts and Crafts Movement was John Ruskin, an art historian and critic teaching at Oxford University. He believed in art as a moral force and idealized the working processes and products made by medieval **guilds.** Ruskin's romantic optimism found its epitome in Gothic architecture.

William Morris was a follower of Ruskin who took a more socialist view of arts and crafts. He believed that happy workers who were not treated like machines inherently made beautiful things. Morris was the most active proponent of the movement in England. He wrote and lectured extensively on the decorative arts and ran an influential design firm.

Charles Robert Ashbee was a follower of Ruskin and Morris and from them developed the idea for a workshop and school. In 1888 he founded the Guild of Handicraft in London. Serving as both a workshop and a school, it took young men off of the streets and taught them various crafts. John Pearson was a trained metalworker and one of the first four workmen. His style of rendering natural forms in copper repoussé greatly influenced the designs of the early metalwork of the guild. Ashbee himself was a self-taught metalworker, and he supplied many of the later designs. Ashbee's designs were characterized by simple, flowing, linear elements, monochrome enamel, and the restrained use of semiprecious stones. The pared down aesthetic of the guild's work may have been tied to the limited skills of its workers, but the minimalist aesthetic is a visual precursor of **art nouveau** and **modernism.**

Inspired by Ruskin, Ashbee was trying to recreate an idealized medieval workshop in which the creative experience of the guild members was valued over the quality of their wares. Unlike the factory workers who each performed a single task repetitively all day, the metalworkers at the Guild of Handicrafts worked single pieces from start to finish, employing all the skills needed to bring a form about. Like many other arts and crafts endeavors, the Guild of Handicraft eventually failed because of the necessarily higher prices that had to be charged for handwork.

Once arts and crafts metalwork gained popularity, its appearance was ironically taken up and copied through industrial processes by larger industries. Metalworker Archibald Knox designed a number of such pieces for Liberty & Co. of London that were made industrially by William H. Haseler in Birmingham. The two most well-known patterns designed by Knox for Liberty were his Cymric line of silver wares and the Tudric line of pewter wares. Both collections showed the influence of **Celtic** motifs that bridged the styles of arts and crafts as well as art nouveau.

The ideals of the Arts and Crafts Movement traveled abroad through publications and lecturers. Much of the arts and crafts metalwork in the United States could be found along the east coast, with Chicago and California also being significant centers of production.

C. R. Ashbee's Guild of Handicrafts helped to inspire the founding of Hull House in Chicago by Jane Addams and Ellen Gates Starr in 1889. Hull House was a settlement house, which strove to improve the lives of the working-class immigrant population in Chicago through education and outreach. Metalwork was taught there for many years with an emphasis on the moral aspects of handwork.

World War II and the cost of handwork brought an end to the high period of the Arts and Crafts Movement, but many of the societies and schools it helped to establish continue to this day. The break with historicism that the Arts and Crafts Movement championed was echoed in the art nouveau styles as well as elsewhere. Arts and crafts beliefs have had an impact on twentieth century industrial design as well as the **Studio Movement.**

See also Art Deco; Bauhaus; Middle Ages; Nineteenth Century; Schwarcz, June

FURTHER READING

The Arts and Crafts Society, http://www.arts-crafts.com

Blakesley, Rosalind P. *The Arts and Crafts Movement.* London: Phaidon, 2006.

Head Hand and Heart. "The Arts and Crafts Movement in Great Britain 1850–1915," http://www.artscrafts.org.uk/index.html

J. R. Burrows and Company. "Founders of the Arts and Crafts Movement," http://www.burrows.com/found.html

Kaplan, Wendy. *The Art That Is Life: The Arts and Crafts Movement in America, 1875–1920.* Boston: Little, Brown, 1987.

Livingston, Karen, and Linda Parry, eds. *International Arts and Crafts.* London: V & A, 2005.

Wissinger, Joanna. *Metalwork and Silver.* San Francisco: Chronicle Books, 1994.

Baroque
(ca. 1580–1730)

"Baroque" is a term used to describe a decorative style that appeared in architecture as well as the fine and decorative arts from the late sixteenth century to the early eighteenth century. The baroque style was dramatic and opulent, and while it dominated Europe, metalwork flourished. It included dynamic and elastic shapes, the inventive use of rich materials, and the imaginative depictions of idealized events. Bold, curvaceous forms that were exuberantly sculptural and ornate characterized much of baroque metalwork. The term "baroque" was used towards the end of the style's popularity and may have come from the Portuguese word for an irregular pearl, or *barroco.*

During the baroque period, the arts were frequently intermingled to the degree that architecture, paintings, and decorative objects within a room were programmatically linked and their boundaries blurred. The increasingly sensual and emotive nature of the baroque also corresponded to the beginning of the empirical sciences, in which sensory experience of phenomena was studied rather than theoretical knowledge.

The theatrical grandeur of the baroque style emerged in Italy from the Counter Reformation as a propagandist tool meant to affect the senses of churchgoers and celebrate the power of the church. Church architecture, interiors, and liturgical objects were created in this style, which came to symbolize Roman Catholicism. The monstrance, used to hold the consecrated Host in Catholic Mass, was particularly a target for elaborate decoration during the baroque period. The style spread from Italy to northern Europe, including England, Germany, Austria, Spain and Portugal. It was disseminated by itinerant goldsmiths, printed illustrations, the Huguenots (see **Paul de Lamerie**), and by overseas missionaries.

The bold drama of the baroque style was intended to have a strong impact on its viewers and was used to aggrandize the positions of the wealthy and powerful outside the church as well. In no place was this achieved more effectively than at King Louis XIV's palace at Versailles outside of Paris. A royal manufactory, Gobeleins, was established to provide a coordinated supply of decorative objects for the palace, including metalwork, tapestries, and sculptures. Many talented artisans were brought in from Italy to execute designs in the fashionable baroque style. The extravagant metalwork in Versailles included solid silver and silver-veneered furniture such as tables, chairs, candelabras, and mirrors.

In the seventeenth century, the fashions of France influenced much of Europe. The new dining rituals begun at the French court called for the invention of new silver tableware as well as the first matching **flatware.** The *service à la française,* in which food was presented artfully on platters at the table, encouraged the development of the tureen, which became a dining table's main feature. French toilette services were also popular, and many were shipped abroad. Unfortunately, much of the French silver from the seventeenth century did not survive because it was melted down to pay for the military campaigns of Louis XIV and Louis XV.

In Germany, Augsburg was a major center of silver production and specialized in tableware, cased tea sets for travel, and silver furniture. Metalwork for local use in Germany was more likely to be **gilt** or partly gilt for increased surface richness. Spanish baroque metalwork was characterized by an increase in the production of damascened wares, particularly in use with ceremonial armor and weaponry (see **Lamination Techniques**). However, the finest Spanish metalwork of the period was executed exclusively for the church in Spain. Much of the baroque wares found in England were collected from abroad when individuals went on grand tours of

THE NEW BEVERAGES

Starting in the seventeenth century, dining and the decorative arts were forever changed by the introduction of three new beverages: tea, coffee, and chocolate. Prior to the introduction of these hot drinks, beverages had been made safer through the inclusion of alcohol. The heating process that was necessary to the making of tea, coffee, and chocolate made them safe as well.

Their introduction brought a revolution in social rituals. Afternoon tea became a regular part of each day in houses that could afford the new beverage. In those same houses, chocolate was served to close friends over breakfast, usually served in the sitting area of the bedroom. Coffee houses opened and became important centers of urban commerce.

The popularity of these and other beverages significantly affected the production of silver wares in the seventeenth century. New forms were developed to accommodate the changing social practices. The teapot, coffee pot, and chocolate pot forms were first developed, but as the social rituals around the consumption of these beverages increased, so did the available equipage for their serving. After acquiring a teapot, one might eventually add to one's service a sugar bowl, cream pot, tea canister, and a waste bowl for used tea leaves. Such a collection of objects was not available in a matching set until the end of the eighteenth century.

Europe. In English manufacture, a somewhat tamed, local baroque style brought a profusion of floral motifs and *putti* (naked, frolicking little boys) on plate forms.

Decorative motifs that were frequently used on baroque metalwork included acanthus leaves, exotic flowers, putti, volutes, friezes, and elaborate scrollwork. In silver especially, elaborate, light catching **chasing and repoussé** was used to add dimension and detail. Pewter frequently had to take the place of silver at the table of the wealthy, when silver was recalled for war funding. Stylistically from this point forward, pewter wares mimicked silver forms. Baroque palaces and churches included fantastic wrought iron gates, balconies, and hardware fashioned to correspond to the overall grandeur.

See also Art Nouveau; Fabergé, Peter Carl; Rococo and Neoclassical

FURTHER READING

Gruber, Alain Charles. *The History of Decorative Arts, Volume 2: Classicism and the Baroque in Europe.* New York: Abbeville Press, 1996.

The Palace of Versailles, Versailles, France, http://www.chateauversailles.fr/en/index.php

Parker, James, Alice M. Zrebiec, Jessie McNab, Clare Le Corbeiller, and Clare Vincent. "French Decorative Arts during the Reign of Louis XIV 1654–1715." *The Metropolitan Museum of Art Bulletin 46,* no. 4 (Spring, 1989).

Seelig, Lorenz. *Silver and Gold: Courtly Splendor from Augsburg.* New York: Prestel Publishers, 1995.

Sorabella, Jean. "Baroque Rome." In *Timeline of Art History*. New York: The Metropolitan Museum of Art, 2000. http://www.metmuseum.org/toah/hd/baro/hd_baro.htm (October 2003)
Walker, Stefanie, and Frederick Hammond, eds. *Life and the Arts in the Baroque Palaces of Rome*. New York: Bard Center, 1999.

Bauhaus
(1919–1933)

Walter Gropius founded the Bauhaus school in 1919 in Weimar, Germany. Although it only existed for 14 years, it has had a dramatic impact on the way metalworking has been taught in much of Europe and North America. Gropius combined the former school of art with the school of arts and crafts to produce *Das Staatliche Bauhaus Weimar*, now known as the Bauhaus. It combined fine arts instruction with a craft apprenticeship system in order to produce prototypes for industry.

Many of the Bauhaus ideas came from the **Arts and Crafts Movement.** Where the Bauhaus differed was in its embrace of industrial practices and functionalist concerns. Functionalism is an approach in which practical issues of use and manufacture play a significant role in an object's design. In the Bauhaus apprenticeship system, students were trained within a specific craft workshop. Both a fine artist and a master technician oversaw their work. This combined method of teaching, to include both the abstractly artistic with the technical, was new to formal education.

At the Bauhaus all students started with the *vorkurs*, an introductory course in which they explored fundamental aspects of design such as line, texture, and color. After the preliminary course, students went on to one of the workshop areas as an apprentice. The workshops included carpentry, stained glass, pottery, metal, weaving, and stage and wall painting. After three years the apprentices could take a test to become journeymen and eventually rise to the rank of master. Johannes Itten, Laszlo Moholy-Nagy, and Marianne Brandt all taught in the Bauhaus metal workshop. Silversmith Christian Dell provided technical leadership.

Objects designed at the Bauhaus metal workshop were meant for everyday life and included tea infusers, ashtrays, and an array of lighting devices. The abstract work of the artists who taught there affected the designs of the objects, resulting in a pared-down aesthetic relying heavily on basic geometric forms.

The political turbulence of Germany in the interwar years presented many problems for such a nontraditional institution. In 1925 the Bauhaus was forced to close and was reopened in Dessau, Germany. In 1928 Hannes Meyer replaced Gropius as director. In 1930 Mies van der Rohe was instated in that position. The school's final move to Berlin was made in 1932, where the Nazi party closed

MARIANNE BRANDT (1893–1983)

Marianne Brandt started as a student at the Bauhaus in 1923 and was the only woman admitted to the metal workshop. Most of the women who applied to the Bauhaus were directed to the weaving workshop. Brandt took over the direction of the metal workshop for a year when Laszlo Moholy-Nagy left in 1928. The lamps and austere geometric tableware that she designed while at the Bauhaus gained her notoriety and were produced industrially under a Bauhaus license. The piece she is probably most well known for is her teapot composed of sections of circles, spheres, and squares. After leaving the Bauhaus, Brandt continued to design for industry and teach.

it in 1933. As the students and staff of the Bauhaus dispersed, they took the ideas of that institution with them. The preliminary course at the Bauhaus has served as a model for the entry-level design courses taken by most people studying art in universities today.

See also Art Deco; Art Nouveau; Guilds; Modernism

FURTHER READING

The Bauhaus Archives, Berlin, http://www.bauhaus.de/english

The Bauhaus Foundation Dessau, http://www.bauhaus-dessau.de/en/index.asp

Bayer, Herbert, Walter Gropius, and Ise Gropius, eds. *Bauhaus, 1919–1928.* New York: Museum of Modern Art, 1938.

Neumann, Eckhard, ed. *Bauhaus and Bauhaus People,* revised edition (Eva Richter and Alba Lorman, trans.). New York: Van Nostrand Reinhold, 1993.

Scheidig, Walther. *Crafts of the Weimar Bauhaus, 1919–1924.* London: Studio Vista, 1967.

Brass. ***See*** Metals

Britannia. ***See*** Metals

Bronze. ***See*** Metals

Byzantine Empire
(330–1453)

When the **ancient Roman** Empire grew too large to govern, it was divided into eastern and western halves. Constantine I, later known as Constantine the

Great, became the ruler of the western Roman Empire in 306 A.D. He defeated his eastern counterpart, Licinius, in 324. Because the eastern part of the empire was stronger than the western part, Constantine moved the capital from Rome to a city known as Byzantium in 330. Byzantium was a city founded as part of ancient Greece in the seventh century B.C. Constantine renamed the city Constantinople, after himself. The western Roman Empire crumbled under attack from without and from that point, the eastern Roman Empire is historically known as the Byzantine Empire for the purpose of clarity. The people of this empire considered themselves Romans, however.

Constantine was the first Christian ruler of the empire, and with him began the delicate balancing act between the inherited Greek and Roman classicism, Christianity, and **Islamic** influence that was to become the hallmark of Byzantine art. These elements were predominantly tempered by an overarching earnestness of religious faith and an aggrandizement of the imperial power that affected all but the most basic and utilitarian crafts. Cultural and religious diversity played out in the metalwork produced in the capital city. Decorative themes included the imperial, mythological, and iconic.

Generosity was considered a virtue of good government, and the emperor made gifts of finely wrought gold and silver to his court, people of Constantinople, and the churches, most especially those in Rome. Gifts were also used as political tools, exchanged between sovereigns, usually passed through their intermediaries. Such gifts of metalwork were wielded to bring attention to the magnificence and prosperity of the Byzantine Empire.

In work for the Christian church, artists and artisans struggled to create powerful devotional icons for worshipers that did not inspire idolatry but rather a direct connection to the spiritual. Sculpture in the round was discouraged for its potential connection to false idols. A controversy over religious icons came to a head in the eighth century. Many icons were destroyed in the hundred years that followed before an understanding was reached between the disagreeing factions about the place of imagery in Christian worship.

After the iconoclastic dispute, greater emphasis was placed on eliminating pictorial elements of a so-called pagan (Greek and Roman) nature. By the eleventh century, a trend toward more elaborate ornamentation and increased figural abstractions developed. Any imagery adapted from non-Christian sources by this time was done without the attached religious references. Decorative elements such as stylized botanical motifs could be borrowed freely from Islamic sources, for instance.

The cultural revival that had occurred between the ninth and twelfth centuries was ended by the crusades and temporary occupation of Constantinople by crusaders. From the twelfth through the fifteenth century, the crafts in general declined as patronage dwindled due to political unrest. The Byzantine Empire ended with the invasion of the Ottoman Turks in 1453. The Turks would eventually change the name of Constantinople to Istanbul, which it is called today.

When Constantinople was established as the capital of the empire, Constantine took great pains to ensure the best craftsmen would follow. Craftsmen were granted special privileges in the city and bound into groups similar to **guild**s or unions. The practitioners of the different crafts each lived in particular parts of town. These discipline-based districts were maintained for hundreds of years. The state regulated the crafts and a hallmarking system was used for gold and silver from the fifth through seventh centuries. The concentration of craftsmen in the capital city was both due to enticements made by the government and because most of the wealthy clientele were there: the imperial court and the church leaders.

SILVER AND GOLD

Household goods retained strong resemblances to Roman forms until the sixth century. The stylization of imagery begun in the later Roman Empire was continued, and figures were represented as slightly abstracted and idealized. This was superimposed with imagery from the new religion, Christianity, juxtaposed with the abstracted vegetal decorations of classical Greece and Rome. Also continuing from the Roman tradition was the practice of the wealthy owning precious metal tablewares. The imperial court and other wealthy members of society owned dishes for feasts, serving dishes, and display plate in silver and gold. In addition to **chasing and repoussé** and **engraving,** it was not uncommon for silver dishes, bowls, and goblets to be inlaid with gold or **niello.**

Byzantine metalwork is most renowned for its religious images. Expensive materials were used to glorify, illustrate, and inspire awe. Liturgical metalwork, objects relating to religious services or worship, made up a considerable quantity of the output of Byzantine goldsmith's workshops. The forms of liturgical objects that had been developed hundreds of years prior to the Byzantine Empire remained largely unchanged. Liturgical products of the metalworker included crosses, chalices, patens, liturgical fans, censers (incense burners), book covers, reliquary or shrine covers, lamps, altar crosses, processional crosses, and icons. These objects would most likely have been wrought in gold, or in silver that was then mercury **gilded.** Their decoration consisted of a number of techniques, generally used in combination. Colored stones or pearls were set at regular intervals, frequently as a framing device. Pictorial imagery was generally represented with cloisonné or champlevé **enamels,** in which stylized figures occupied flat spaces, arranged in a grid format. In the portrayal of such figures as the Virgin Mary, Christ, and saints, guidelines were developed regarding their placement and position in a given context. All of these parameters were intended to create a universal visual language for the devout.

In the eleventh century the eastern (Byzantine) and western (Roman) Christian churches split. What was the eastern Christian church in Byzantium is known today as the Eastern Orthodox Church. Icons were an important part of the Eastern Orthodox Church, and they were made in all media and sizes.

An icon is a representation of the Virgin Mary, Christ, or saints that is meant to serve as an intermediary to the religious figure referenced, to which petitions were put in the form of prayers in hope that the personage would intercede on the devout person's behalf. Metalwork was originally used to frame the icons, but eventually gold work encroached into the background of the image itself. These representations became more like frozen theatrical moments over time, encouraging viewer empathy and emotional participation.

The richness of the decoration also continued to escalate, both in materials and design motifs. Stones, cameos, and enamels were added where before had only been chased metal. Finely worked **filigree** was intertwined to form organizational subdivisions. The bowls of some chalices were carved from hard stones such as jasper and mounted on elaborate gold stems and bases.

Within this splendor, no object was more lavishly decorated than the reliquary. A reliquary is a container for an object of spiritual value such as the remains of a saint or a piece of the True Cross on which Christ was crucified. These came in many forms, from the architectural to the corporeal. Many examples of reliquaries for the True Cross were in the form of a box with a sliding lid. The exterior generally included images of the crucifixion and other scenes from the life of Christ. These and other reliquaries were venerated by the faithful and were sometimes the cause of religious pilgrimages.

BRONZE

Bronze continued to be used in households or smaller churches that could not afford silver. Although the making of bronze statuary declined during the Byzantine Empire, **casting** of massive bronze doors for churches continued this sculptural tradition. These massive architectural details were made for churches and ornamented with silver **inlay.**

ENAMELS

Enamel was used to decorate increasing quantities of objects as time went on. Enamel, which had been used only by the so-called barbarians living on the periphery of the Roman Empire during its peak, was now adopted for the glorification of the Christian religion. Cloisonné was very popular, especially in liturgical pieces that relied on symbolic representation for communication. By the eleventh century, fields of gold were surrounding the cloisonné.

See also China; Persia; Verdun, Nicholas of

FURTHER READING

Beckwith, John. *Early Christian and Byzantine Art,* revised edition. 1980. Reprint. Yale, 1986.
Dumbarton Oaks, Research Library and Collections, Washington, DC, http://www.doaks.org
Evans, Helen C., and W. D. Wixom, eds. *The Glory of Byzantium.* Metropolitan Museum, 1997.
The Hermitage Museum, St. Petersburg, Russia, http://www.hermitagemuseum.org

Lowden, John. *Early Christian and Byzantine Art.* Phaidon, 1997.
Rodley, Lyn. *Byzantine Art and Architecture.* Cambridge, 1994.
Saint Mark's Basilica, Venice, Italy, http://www.basilicasanmarco.it/eng/index.bsm

Casting

The process of casting involves heating metal until is reaches a liquid state, then guiding it to a negative space where it cools and hardens back into a solid. The resulting object is referred to as a cast or casting. By controlling the shape of the negative space into which the metal is poured, known as a mold, almost limitless forms may be achieved. Rudimentary metal casting was being done in Egypt and Sumer as early as 3500 B.C.

The earliest castings were most likely done with open-faced molds that were carved recesses in stone or clay. Molds continued to develop into multipart forms that would fit together to allow the creation of objects in the round. In an enclosed mold such as that, a channel must be created in which to pour the molten metal. A channel used to deliver molten metal to a mold cavity is called a sprue.

Any substance that will withstand the heat of the metal being poured into it may be used as a mold. The mold material might be carved (such as with stone), molded (such as with clay), or poured and allowed to harden (such as with plaster). Early cultures tended to use whatever materials were most readily available to them. In pre-Columbian South and Central America, soft stone molds were used. European coastal regions may have preferred the dried, cleaned bone of the cuttlefish. In certain areas, sand deposits with low clay content were found, which allowed for sand casting.

Molds may be said to be of two kinds: waste molds and reusable molds. Waste molds are destroyed or wasted during the process of casting. Reusable molds, such as those made of steel, are stronger and maintain their structural integrity over many castings.

SAND CASTING

Traditional sand casting does not allow for complex shapes with undercuts, but it is a very economical means of casting. In sand casting, a replica of the object to be cast, called the pattern, is made. The pattern must be made of a material durable enough to repeatedly withstand the mold-making process, such as metal or wood. The pattern is then placed in a flask, which is a framelike four-sided box. The flask is then firmly packed to the top with sand to cover the pattern. The flask is turned over and a second flask is placed on top and filled with sand.

The lower flask is known as the drag, and the top portion is called a cope. Once packed sand surrounds the pattern, it is removed, and parting dust, such as charcoal, is sprinkled on the seam between parts to help with separation later. A sprue is cut into the cope to allow an entry point for the metal. More complex forms may also call for a gate, which is a passageway similar to a sprue that lets air escape a mold as metal enters. Once both parts of the flask are reassembled, metal may be melted and poured into place. With this kind of traditional sand casting, the mold is destroyed each time, but the sand and the pattern may be re-used.

Today, sand bonded with resins and other additives is frequently used. It is a common practice in the automotive industry to cast engine blocks by packing a carved foam pattern in sand, then pouring the molten metal into the foam. This burns away the foam original while replacing it with metal.

LOST WAX CASTING

The technique of lost wax casting was developed by the third millennium B.C. Although there are many permutations of this technique, in all instances there is a pattern of the final object made in wax that is destroyed with heat. Taking a mold off of a pattern in another material, such as clay, and then pouring enough wax in to the negative space to form a shell in the same shape as the original pattern is one way to create the wax pattern. The wax might also be formed directly with tools. Casting metal very thickly tends to lead to cracking of the metal as it cools, therefore large pieces are cast with hollow cores.

In traditional lost wax lasting, the wax pattern is fitted with wax wires and rods that will serve as sprues and gates. This object is then placed in a flask; this may be a section of pipe or may simply be a tube reinforced with wire. The space around the wax is filled with a plaster-like refractory substance called investment. The term refractory means that it can withstand high temperatures. The investment dries, the wax is melted out, and while the mold is warm, molten metal is poured in. Hollow core casting may be achieved by either modeling the wax over a refractory core or by using a wax shell that is both surrounded by and filled with investment. In either case, pins are frequently used to suspend the core so it does not shift during the wax burnout.

Contemporarily, it is more common to see artwork cast through the lost wax process using ceramic shell instead of investment. Ceramic shell is a refractory substance that is built up around the wax pattern through the application of a liquid known as slurry. While the slurry is wet, a grainlike substance known as stucco is applied. Once dried, the steps are repeated until a shell of the desired thickness in achieved. In each application, the size of the stucco granules increases.

SLUSH CASTING

In slush casting, the metal is poured into the mold to coat the walls, then a portion that is still liquid is poured back out, creating a shell. This leaves a hollow shell of metal and is done to conserve metal.

OTHER TERMINOLOGY

Die casting is an industrial process that uses pressure to force molten metal into reusable metal dies. When molten metal is guided to a mold simply by gravity it is referred to as a gravity pour or gravity casting. Metal may also be forced into place with suction, as with vacuum casting. Centrifugal force and steam may also be used to ensure that metal has filled a mold. The place where casting takes place is referred to as a foundry. The container in which metal is melted is called a crucible.

See also China; First Metalsmiths

FURTHER READING

Ammen, C. W. *Metalcasting.* New York: McGraw-Hill, 2000.
Langland, Harold. *From Clay to Bronze: A Studio Guide to Figurative Sculpture.* New York: Watson-Guptill Publications, 1999.
Tolles, Thayer. "American Bronze Casting." In *Timeline of Art History.* New York: The Metropolitan Museum of Art, 2000. http://www.metmuseum.org/toah/hd/abrc/hd_abrc.htm

Cellini, Benvenuto
(1500–1571)

Benvenuto Cellini is one of the most well-known goldsmiths in the history of Western decorative arts. Although only one of his pieces in gold is known to have survived, his works in other media and autobiography have made him an illustrative figure of his times.

Cellini was born in Florence, Italy, to a father who had made his living in music performance and architecture. Cellini began his goldsmith's training at the age of 13 and worked in various goldsmiths' shops in northern Italy until the age of 19. From 1519 to 1540 he moved around a great deal, predominantly working in Florence and Rome. While in Rome, Cellini made several pieces for Pope Clement VII including dies for the striking of coins in the papal mint and portrait medals. Living in Rome also enabled Cellini to study the work of such masters as Michelangelo and Raphael.

In 1540 Cellini's ambitions led him to France and the court of François I at the palace of Fountainbleau. During the five years he was there, Cellini received many commissions from court members for smaller personal luxury items. He was also asked to design several architectural elements for the palace, including a bronze wall relief known as The Nymph of Fountainbleau. This was Cellini's first experience with casting bronze on a large scale. Indeed, the many substantial sculptures that Cellini executed while in France established his reputation as a sculptor as well as a goldsmith.

It was while at the court of François I that Cellini made his now famous saltcellar. Although it was originally designed for Cardinal Ippoito d'Este, it was finally executed for François I. This is arguably the most outstanding work made by any goldsmith during the **Renaissance,** the period of revived interest in learning and the arts that spread through Europe starting in the fourteenth century. Many examples of the goldsmith's trade have not survived because they were melted down and converted to currency or remade in keeping with the changing fashions. Even without many other examples of work by his contemporaries, the saltcellar by Benvenuto Cellini can be understood as a masterpiece of the goldsmith's art.

SALTS AND SHIPS

A saltcellar, sometimes called a salt, is a decorative container for salt and sometimes pepper. Salt has had enormous symbolic importance at various times in history because it is essential to the human diet and widely used as a preservative. In **ancient Rome,** soldiers were given an allowance for salt known as salarium, from which the word salary is derived. From the **Middle Ages** until the seventeenth century the very wealthy typically had elaborately sculpted saltcellars at their tables. Salt was a relatively costly seasoning at the time and therefore was presented in a manner befitting its status at the table. Kings had elaborate salt containers placed near them while they ate. Those who were seated further down the table were said to be "below the salt," which was an indication of their lesser importance. The form of Benvenuto Cellini's famous saltcellar combines the traditional form of a salt with a nef. A nef is a small model ship made of costly metals that sat on a dinner table to mark the location of the host and sometimes served as a container for his personal **flatware** or salt.

Cellini's saltcellar is executed in gold and **enamel** on an ebony base. The overall form is oval and stands only ten inches high. The two major elements are figures that represent the sea and the earth in the manner of classical mythology. Respectively, they embody the sources of salt and pepper. The figure of the sea is surrounded by sea creatures and holds a trident. At his right hand is a small boat that was meant to be a receptacle for salt. The figure of the earth rests on the ground with a miniature temple at her right hand in which pepper was kept. Around the base are figures in relief representing the four parts of the day: morning, midday, evening, and night. These four are nearly identical to ones created on a large scale by Michelangelo in Florence.

The personifications of sea and earth were **raised** from sheets of gold while the smaller objects, such as the temple, were **cast. Chasing and repoussé** are also evident in the contours of the land and ocean that the figures rest on.

This saltcellar is considered to be mannerist in style because of the elongation and posturing of the figures as well as its heavy ornamentation. Mannerism was an Italian courtly style developed in the first third of the sixteenth century that was used extensively at the palace of Fontainebleau in France. Its characteristics include exaggerated human forms, complex ornament, and references to the natural world as well as an emphasis on useless display.

After leaving Fountainbleau, Cellini returned to Florence, where he continued to work as a goldsmith as well as a sculptor in bronze and marble. Around 1558 he began writing his autobiography, the *Vita di Benvenuto Cellini.* Cellini's decision to write his own biography was rooted in the humanistic belief of the Renaissance that those who led an extraordinary life were responsible for documenting it. Although not always strictly factual, the *Vita* does illuminate Cellini's turbulent personal life, including some of the conflicts that caused his numerous relocations and many unfinished commissions. It also describes many of the pieces that he produced during his lifetime that have since been lost. Furthermore, Cellini's account of his own life has been a major source of information about techniques used by goldsmiths and sculptors during the Renaissance.

FURTHER READING

Kunsthistorisches Museum, Vienna, http://www.khm.at/homeE/homeE.html

Pope-Hennessy, Sir John Wyndham, David Finn, and Takashi Okamura. *Cellini.* New York: Abbeville Press, 1985.

Tamarin, Alfred H., John Addington Symonds, and Benvenuto Cellini. *The Autobiography of Benvenuto Cellini.* New York: Macmillan, 1969.

Celtic People
(ca. 700 B.C.–1400 A.D.)

It is believed that Western European settlers became what we know as the Celtic people through general accumulation during the first millennium B.C. They were a warring people who spread across much of Western Europe from the eighth century B.C. to the first century A.D. It was the Greeks and Romans who called these people Celts. The Celtic people were made up of many smaller tribes or kingdoms that were all linked by language and culture. At the time of their greatest range, in the third century B.C., the Celts occupied lands from the British Isles in the west, south to the Iberian Peninsula, and east to Asia Minor.

Celtic social strata included the warring nobility, a learned class of druids, writers, and dispute mediators, and the lower class. There were no urban centers and few crops were raised. Celtic life revolved around cattle herding and fighting. The metalworking that the Celts practiced reflected this lifestyle. Much of what has been found in burial sites has included weaponry, horse trappings, and chariot hardware. They were early specialists in ironwork, which appears to have flourished under royal patronage.

Most of the Celtic people were absorbed into the Roman Empire beginning in the first century B.C. Previously Celtic areas became a cultural blend of Roman and Celtic influences. In the areas where the Roman Empire did not reach, such as present day Ireland, Scotland, Wales, Cornwall, the Isle of Man, and northwest France, Celtic culture persisted, and variations of the Celtic language continue to be used in those locations to this day. This was a great reduction in size from the original expanse of Celtic influence and is sometimes referred to as insular Celtic.

When Christianity came to Ireland in the fifth century, Celtic metalwork was used in the service of the new religious practices, creating a unique visual language. Following the introduction of Christianity and the written texts that accompanied it, metal workers in Celtic regions began to produce elaborately worked book covers, influenced by Anglo Saxon work of this kind.

Two historic periods in Celtic Europe were the Hallstatt and La Tène. Hallstatt is the name given to a late Bronze and early Iron Age culture that was predominant in central Europe for 200 years from about the eighth century B.C. The name is derived from the location near Salzburg, Austria, at which large quantities of objects typical of the period were discovered. Some of the earliest evidence of the Celtic people comes from this time.

During this stage, bronze work became more sophisticated, incorporating items worked from flat sheet as well as those that were cast into their final form. The use of silver and iron increased. Some articles have embossed gold decorations (see **Stamping and Die Forming**). The geometric decorative motifs found from this period may have been the result of contact with the classical world. It was a time of great technical improvements in metalworking, which the Celts would take with them as they conquered new lands.

Hallstatt was followed by the La Tène Period, named for an excavated site on the shores of Lake Neuchâtel, Switzerland. It encompassed the late Iron Age Celts and was roughly contemporaneous with **ancient Greece.** The La Tène period is considered by some to be the high point of Celtic culture, lasting from around 500 B.C. to the Roman invasions in the first century A.D. It is at this time that the Celtic people begin to use a silver coinage system based on those of the Greeks and Romans.

This is the first period with significant evidence of lost wax **casting** in Western Europe. Previously, they had used mostly bivalve ceramic molds, possibly due to lack of domesticated honeybees. Lavish display pieces included colorful elements such as semiprecious stones and coral. A kind of red **enamel** that was held in place with cold **connections** rather than by fusing to the metal was also used. Thin gold foil and **filigree** were affixed to the surface of prestigious objects. In addition to personal decorative objects, metalworkers of the period produced mirrors, hanging bowls, and vessels for domestic use.

In the middle of the first millennium B.C. the Celtic regions of Ireland, Scotland, north Britain, and Wales experienced the Iron Age. Iron was produced in significant quantities to make it available to most levels of society. In addition to swords and weapons, iron implements for the planting and reaping of grains, as well as axes and saws, can be found form this period.

The decorative style of La Tène was dominated by symmetrical spiral and circular geometric designs that were interlaced. Abstracted plants and animals were also arranged into stylized designs on heavily decorated surfaces. The La Tène period influenced medieval art in the few remaining Celtic areas of northwest Europe (see **Middle Ages**).

In general, Celtic metalwork is best known for its geometric design motifs that frequently included stylized beasts, people, and floral elements within them. The layout for designs was planned on grids drawn on bone or slate prior to working in metal. The circular motifs were planned with compasses. These elements were initially shallow but became deeper over time.

The technical abilities of the ancient Celts included **engraving,** casting, **etching,** and overlays of thin metal cladding in bronze and gold. Coral, glass paste, shell, **niello,** and amber were used for color. They also plated with tin, gold, bronze, and a silver and copper alloy (see **Gilding and Plating**). Plating and **inlaid** materials seem to have been particularly used to create contrasting surfaces. The glass used was probably not true enamel but rather a glass paste that was pushed into recesses after being softened by heat. The glass was mixed with cuprous oxide to give it an opaque red color. Patterns in **chasing and repoussé** in bronze and brass were found on shield bosses and scabbards. Small pieces in relief were created in large quantities through an embossing system.

Medieval Celtic metalwork became increasingly sophisticated. By the seventh century, it incorporated a full range of bronze- and blacksmithing, with most skills translated into silver as well. The ancient Celts had generally used silver alloyed with gold. Silver was only used by itself much later. True enameling was used in an increasing range of colors. Decorative elements included linear ribbing, crosses, textured backgrounds, triskelions (tri-cornered spirals) and peltae. A pelta is a design structurally based on two equal-sized circles fitting exactly next to each other and within a larger circle. Stamped geometric designs were also incorporated in repetition. In the medieval period, decoration was once again made shallow. Larger pieces were composed of smaller elements and usually attached with cold connections.

Decorative book covers contained gridded panels of contrasting decoration. In the Middle Ages, Celtic nobility used highly decorative horse bridals, incorporating enamels, **gilding,** and tinning. In the eighth and ninth centuries, the metal fittings for drinking horns were commonly cast from bronze in the forms of bird or dragon heads. With the arrival of Christianity to places such as Ireland came a greater need for Eucharist objects. Monasteries increasingly became centers of metalworking.

BRONZE

As a primarily warring society, the early Celts reserved most of their technology for **armor and arms.** Bronze was used for most weapons and tools prior to the rise of iron. It was used to make hardware for leather or wooden shields, swords, scabbards, knives, daggers, and horse fittings. A few domestic objects found from this period include cauldrons, mirrors, and wooden vessels decorated or held together by bronze.

IRON

Unlike its European predecessors, Celtic culture used iron in significant quantities. By approximately 600 B.C. iron was becoming a more popular metal than bronze for tools and weapons. Early iron mimicked bronze forms at first.

Celtic graves show swords and wagon fittings in iron created for wealthy patrons. Having control of proficient blacksmiths was a mark of affluence and power and the nobility gave gifts of metalwork to show their status. Weapon makers frequently signed their best work. It is probable that smiths outside the system of royal patronage existed to make simpler utilitarian objects and tools. Blacksmiths also made firedogs and chains for cauldrons. Advancements in ironworking technologies underpinned many advances in the Celtic civilizations.

Periodically, bronze and iron were combined. One way of connecting iron and bronze was to physically cast the metal with the lower melting temperature, bronze, around the metal with the higher melting temperature, iron. This technique, called casting on, was especially found in horse and vehicle hardware. Items have also been found in which the iron object was heated and dipped in molten bronze in order to plate the iron. Bronze and iron were used together in some La Tène cauldrons for improved durability. Bronze was also inlaid into some iron weapons as an embellishment.

HANGING BOWLS

Some of the great mysteries of the Celtic people are the hanging bowls that they have left behind. It is not known what these were used for. Dating from the fifth to the eleventh centuries, they were made of beaten copper-based alloys.

These bowls appear to have been suspended from three or four hooks. Mounts for the hooks, called escutcheons, were located outside and below the rim. They were decorated with openwork, enamel, or cast into the forms of creatures, and soldered or riveted on. Some bowls may have been polished on a **lathe.**

The objects that survive from the ancient Celtic people are primarily functional. Their geometric motifs continue to be popular to this day and were notably incorporated into the silver and pewter of Archibald Knox in the nineteenth century (see **Arts and Crafts Movement**).

See also Byzantine Empire; First Metalsmiths

FURTHER READING

The British Museum, London, England, http://www.britishmuseum.org

Chadwick, Nora K., and Barry W. Cunliffe. *The Celts.* London: Penguin Books, 1997.

Green, Miranda J. *The Celtic World.* London: Routledge, 1996.

The Hueneburg Museum, Hundersingen, Germany, http://www.dhm.de/museen/heuneburg/indexe.html

Laing, Lloyd Robert. *A Catalogue of Celtic Ornamental Metalwork in the British Isles, c. AD 400–1200* (BAR British series, 229). Oxford: Tempvs Reparatvm, 1993.

MacGregor, Morna. *Early Celtic Art in North Britain: A Study of Decorative Metalwork from the Third Century B.C. to the Third Century A.D.* Leicester, England: Leicester University Press, 1976.

Stead, I. M. *Celtic Art: In Britain before the Roman Conquest.* Cambridge, MA: Harvard University Press, 1996.

Youngs, Susan, and P. T. Craddock. *The Work of Angels: Masterpieces of Celtic Metalwork, 6th–9th Centuries AD.* Austin: University of Texas Press, 1990.

Chasing and Repoussé

The techniques of chasing and repoussé are used to give decorative form to planar metal by stretching or compressing the material. Although they may be utilized separately, these two techniques are frequently used in conjunction to achieve ornamental effects in relief. The resulting work is described as chasing or repoussé work, respectively. Chasing and repoussé work has been found dating back to at least 1600 B.C.

Chasing may be done to create indentions or impressions into the surface of a sheet of metal, usually on the front side. It may also be utilized to sharpen detail on a cast metal object. Repoussé adds volume and decorative relief by stretching the metal from behind.

Both processes are performed on a smaller scale with punches struck with hammers, or on a larger scale with direct hammer blows. The metal is generally worked when in an **annealed** state. The small punches are usually held at an angle against the metal and tapped with a chasing hammer, which resembles a small ball peen. As the hammer strikes the punch, the punch is guided by hand along the surface of the metal to create small, overlapping indentations. The metal is worked on one side, then the other, and the desired form slowly develops through repeated courses. When the two techniques are used together, chasing is generally done first to outline the design. Flat chasing is the use of chasing techniques by themselves to impress designs that are usually of a linear nature.

While the metal is being pushed from one side, it is usually being supported on the other by a substance that yields minimally. Historically this has included lead, leather, wax, and wood. Contemporarily, most chasing and repoussé work is done with the use of pitch as support. This is a tarlike mixture consisting mainly of a kind of asphalt. It comes in varying degrees of hardness and can be melted at relatively low temperatures.

FURTHER READING

Lewis, Marcia. *Chasing: Ancient Metalworking Technique with Modern Applications,* second edition. Long Beach, CA: LarMar Productions, 2000.

Steines, Adolf. *Moving Metal: The Art of Chasing and Repoussé.* Huntingdon, PA: Blue Moon Press, 2001.

China

Although this entry will discuss metalworking from the geographic region currently known as China, that area was not a unified cultural entity for all of its recorded history. Although generalizations about China's history of metal use and production may be made, they must not be taken as universally true for the whole of the current country.

The earliest methods of working with metal in China were most likely developed autonomously. The smelting and **casting** methods needed to create objects of bronze and iron were mastered early and reached high levels of technical expertise. Unlike Western civilizations that prized gold and silver above other materials, the most culturally important metal in ancient China was bronze. It was not until the importation of Buddhism that gold and silver objects gained significance. **Enameling** techniques were also introduced later in China's history but were quickly adapted and used in the creation of splendid objects.

BRONZE

Bronze has been cast in China since the second millennium B.C. At the time it was the most advanced technology of the day. Chinese bronzes usually include lead in addition to the standard tin and copper, which sometimes results in a grayish appearance.

Unlike most early cultures that used the lost wax method of casting, the earliest Chinese bronzes were made with piece molds. A model of the object to be cast would be made of clay or another workable material. Once finished, a second layer of clay would be molded around the original and, once dried, removed in sections. The outer layer of clay took with it the impression of the object to be cast and became the mold. Additional decorations could be incised right into the mold's interior prior to firing it. Once the fired parts of the mold were reassembled they provided a negative space that molten metal could be poured into and that would withstand the high temperatures of casting. When the metal had cooled, the mold could be removed from the cast object for future use. Because most bronzes were cast in series, the mold could not contain any undercuts that would prevent their removal once the bronze was cast. Thus in the ancient Chinese bronzes it is common to see very straightforward, low relief as the means of decoration. The lost wax process of casting and the hammer forming of sheet metal came into use later.

The production of cast bronze articles was well organized and consistently yielded large quantities of high-quality wares. As with modern assembly lines, each aspect of the bronze casting was subdivided into specialized tasks for many laborers. The ruling class controlled production. The objects cast by the early Chinese included weapons, chariot hardware, and ritual vessels. In later periods, other objects such as mirrors, garment hooks, and ladles were also cast. The cast ritual vessels mimicked ceramic vessels of the period, which were made in prescribed forms according to their functions. These included holding wine, cooking and holding food, and ritual cleansing.

Bronze vessels were not used to cook ordinary meals but for rituals and to make offerings to ancestors. It was believed that a person's ancestors had the ability to intercede with the gods on behalf of living family members. All family groups also made offerings to their ruler's ancestors because society was seen as a macrocosm of the family structure.

Chinese society was dependent on clear hierarchies. Ritual vessels for the worshipping of one's ancestors were designed to reinforce these hierarchies and make one's position in society clear. The more important a person was within the stratified society, the larger, finer, and more numerous that person's bronzes were. Because bronze foundries were controlled by the state, wealth alone was not enough to acquire more impressive vessels. Ritual vessels were also buried with the dead in order that they might continue to make offerings in the afterlife. Tombs have been found containing as many as 200 bronze vessels.

Inscriptions included on the cast bronzes constitute some of the earliest records of Chinese calligraphy. From the sixteenth century B.C. to the eleventh century B.C., the inscriptions evolve from simple pictograms to long inscriptions including valuable historical information such as dates and events.

Decorative motifs on these vessels frequently incorporated animal forms, including dragons, birds, snakes, and cicadas. One of the main relief motifs on these vessels was an animalistic face known as the *taotie.* It is not known what these images represented, but they usually consisted of a masklike arrangement of eyes, ears, mouth, and horns, with the possible addition of claws and other body parts. Abstract geometric motifs were also found in horizontal zones on some vessels. As time passed, the appearance and importance of the ritual vessels changed. The imagery became more secular, and around 400 B.C. the Chinese began to **inlay** gold, silver, and stones in cast bronze.

Buddhism arrived in China after the first century A.D., bringing with it a major influx of foreign ideas. Instead of focusing on the familial and hierarchical relationships of society, Buddhism emphasized personal salvation. At this point, cast statuary and other devotional implements of Buddhism began to be made in significant quantities. As a country that is very much in touch with its past, however, the Chinese continued to study and create bronzes based on the ancient examples.

IRON

The Chinese were the first people to cast iron, many hundreds of years in advance of the West. By the fifth century B.C., the mining and smelting of iron had been achieved in the southern regions of China, and the technologies were traveling north. It is likely that the techniques of smelting and casting iron developed as an outgrowth of the methods used to process the ingredients for bronze. Because iron processing requires much greater heat than the components used in bronze, most societies were not able to work with mined iron until much later.

Iron was used for weapons, tools, and hardware. Methods of mass production were developed that ensured the swift manufacture of many standardized, identical elements. These methods of subdivided labor that were used very early on in China would eventually have a great impact on other countries. Today, China is one of the world's leading producers of iron ore.

GOLD AND SILVER

Before the seventh century A.D., gold and silver were mostly used for creating decorative inlaid patterns in bronze and lacquer wares. Those two metals that were so prized by other cultures were almost never included in early Chinese burial rites. In the third through sixth centuries A.D., examples of silver objects from southwestern Asia and the Mediterranean made their way into

China under foreign rule. Included were forms that had previously been unknown in China, including cups with stems and lobed bowls. These and other new shapes were adopted by Chinese metalsmiths and modified to appeal to regional preferences.

In central and southwestern Asia some of the most treasured sculptures and reliquaries associated with Buddhism were made in gold and silver. Thus, when that religion gained followers in China in the fourth and fifth centuries A.D., it brought an increase in the number of religious-themed gold and silver work. From the fourteenth through the early twentieth centuries, silver and gold cups and plates could be found at court, and their use in religious rites was generally restricted to the upper nobility.

ENAMEL

The technique of enameling on metal may have appeared in China as early as the seventh century A.D. The cloisonné technique of enameling, which is typically associated with this culture, arrived in the fourteenth century and was most likely brought from **Byzantium.** Chinese cloisonné was done with copper or bronze wires on copper or bronze vessels. After the spaces created by the wires were filled with enamel, the top ridge of the wire and any other exposed metal parts were frequently **gilded.** The bright colors achieved through enameling were originally considered garish and only appropriate for women's articles such as boxes or flower vases. The decorative motifs used were derived from existing Chinese imagery found on ceramics, lacquer ware, and bronzes.

By the fifteenth century, cloisonné articles were commonly made for the royal palace and included large vessels such as fish bowls and washbasins. A height of technical virtuosity in cloisonné enameling was reached in the eighteenth century, and Chinese enamels were made in great number for export. The Chinese practiced champlevé enameling at the same time they were working in cloisonné.

The painterly use of enamels that was perfected in Limoges, France, appeared in China in the seventeenth century. The technique was used much like china painting on ceramics but applied to prepared metals. The imagery used in painted enamels followed the trends of similar porcelain forms, beginning with traditionally Chinese motifs and becoming gradually Westernized.

See also Flatware; India; Japan

FURTHER READING

Art Museum at the Chinese University of Hong Kong, China, http://www.cuhk.edu.hk/ics/amm/

Asia Society Museum, New York, http://www.asiasocietymuseum.com

Burling, Judith, and Arthur Hart Burling. *Chinese Art.* New York: Bonanza Books, 1953.

Feddersen, Martin. *Chinese Decorative Art.* New York: Thomas Yoseloff, 1961.

Freer Gallery of Art. *Masterpieces of Chinese and Japanese Art: Freer Gallery of Art Handbook.* Washington, DC: Smithsonian Institution, 1976.
Freer Gallery of Art and Arthur M. Sackler Gallery, Smithsonian Institution, Washington, DC, http://www.asia.si.edu
Hong Kong Museum of Art, Hong Kong, China, http://www.lcsd.gov.hk/CE/Museum/Arts/english/intro/eintro.html
Palace Museum, Beijing, China, http://www.dpm.org.cn/english/default.asp
Rawson, Jessica, ed. *The British Museum Book of Chinese Art.* New York: Thames and Hudson, 1993.
Shanghai Museum, Shanghai, China, http://www.shanghaimuseum.net
Tregear, Mary. *Chinese Art.* New York: Oxford University Press, 1980.
Wagner, Donald B. *Iron and Steel in Ancient China.* New York: E. J. Brill, 1993.

Colonial America
(ca. 1600–1775)

Metal goods in Colonial America were an integral part of life, from necessities to luxuries. Although the colonies tended to follow England stylistically, their particular circumstances created both special problems and solutions. Encouraged by the riches found by the Spanish in Central and South America, England and other countries sent colonists to the New World. They hoped to find resources that could be sent back for manufacturing, such as metals. The colonies also served as a captive market for manufactured goods.

Workers in the metal trades were among the first settlers in America. They were needed for their technical abilities in the repair and manufacture of tools but also because metals were among the riches that colonists hoped to find in America. As the population grew, so too did the need for workers in the metal trades, including silversmiths, coppersmiths, braziers, and blacksmiths. By the end of the eighteenth century there was considerable specialization within the trades.

Although deposits of copper and iron were found in the colonies, other base metals were absent, causing colonists to be dependent on England for finished objects in these metals. The large-scale industries in England also proved difficult competition for the newer, smaller enterprises in the colonies.

Stylistic influences were acquired through the training of apprentices (see **Guilds**). They also derived from printed advertisements as well as extant examples. Furthermore, regional styles were influenced by the nationalistic preferences of the immigrants who settled America including the Dutch, German, and French.

GORGETS AND GIFTS

One of the last elements of a full suit of plate **armor** to be worn with a military uniform was the gorget—a curved plate of metal that had originally protected the neck of the wearer. As warfare changed and the use of metal armor fell by the wayside, the gorget became a decorative element worn at the neck by those in uniform, frequently wrought in silver. It served as a reminder of the days of knights in shining armor.

Officers in the British military had them on their uniforms during the colonization of America. Somehow, during the settlement of North America, Native Americans came to actively prize and trade goods for the gorgets themselves. Eventually the colonists made gorgets for the specific purpose of trading with or giving to the Native Americans.

SILVERSMITHS

Silversmiths were at the Jamestown settlement in North America as early as 1620, but it would be another 20 years before there was enough silver in the colonies to necessitate their services. Early silversmiths generally did a lot of repair work rather than manufacture new items. They also remade old articles to suit the latest fashions. Because raw materials from England were highly taxed, colonists' silver tableware was usually ordered from England or made by melting down silver coins locally. Having one's coins made into silver tableware was a practical use of wealth. It could be used to eat or drink with, to show off one's wealth, sold to pay debts or inheritance, and was far easier to identify if stolen than a pile of coins.

Although the first silversmiths in North America were trained under the English guild system, there was no such authority in the New World. It was therefore of even greater importance that the silversmiths be honest, from a respected family, and with a good reputation. With the exception of a brief period in Baltimore, there were no assay offices in the colonies to check that the silver standard was being maintained. Even without a hallmarking system, silver was still marked by the workshop in which it was made. Silversmiths were generally in the upper class of tradesmen because of their prestigious materials and wealthy patrons.

The dominant styles in silver followed English examples. This was especially true in towns like Boston that were primarily settled by the English. New York had a number of Dutch immigrants, and a number of Huguenots settled there later, affecting the stylistic traits of local silver (see **Paul de Lamerie**). In a similar manner, Pennsylvania metalwork exhibited influences from the German immigrants who settled there.

The upper-class colonists of the southern regions were more philosophically aligned with England. They were therefore more likely to order their silver from

London merchants, paying the cost with the profits from exported agricultural goods such as tobacco. There were a few itinerant silversmiths in the south, but most were indentured servants or slaves that served on plantations.

It was the man of the household who usually selected and owned silver objects. The first pieces of silver that a person generally bought were spoons, followed by a cream pot. Tea wares were purchased individually, as sets were not available until after the revolution. Other household items made of silver included forks, communal and individual drinking vessels, saltcellars, and punch bowls. As the country grew, so did the list of available items. Major centers of silver production developed in Boston, New York City, Baltimore, and Philadelphia.

Silver was also given as gifts on the occasions of birth, marriage, and death. In New York it was common to give spoons at funerals, christenings, and some birthdays. Churches that could afford it had silver for communion and baptism, usually given by members or purchased with money given by the congregation for that purpose.

Because raw materials were not easy to come by in the colonies, silversmiths developed a number of material-saving techniques, including the use of thinner metal and less solder. Otherwise, the methods used by colonial silversmiths did not differ radically from their European counterparts. Most early vessels were made by **raising.** Additional elements were produced through sand **casting** and attached by soldering (see **Connections**). Silver was drawn through shaped metal plates to make banding that could be soldered to rims and edges. **Chasing and repoussé** and **engraving** were also used. In the eighteenth century, water- or steam-powered rolling mills made prepared sheet metal more widely available, leading to pieced sheet work.

Most silversmiths practiced some kind of side business. In addition to making silver objects they might also retail imported ones. If they were skilled engravers they might be called on to cut copperplates for printed matter. They also engraved the first colonial currencies and minted the first coins. They were additionally called on to ornament guns or sword hilts. Sidelines for silversmiths that were less like their trade included dentistry and inn keeping (see **Paul Revere**).

COPPER AND BRASS

The term coppersmith generally denotes a professional who makes objects from copper. The term brazier is used to describe one who works with brass. Stylistic changes in copper and brass were contemporaneous with their precious metal counterparts, though frequently less decorated.

Although copper was mined in America as early as 1709, the majority of copper used by colonists came from imported English sheet copper. Copper objects were made both by raising and by seaming. In seaming, sheets of copper were fitted together with tabbed seams and reinforced with a low temperature solder. Common goods made from copper included stills for alcohol, teakettles, chocolate pots, and fish kettles. Household vessels that came into contact with food were coated with tin on their interior to prevent the copper from reacting with the food.

Made of copper and zinc, brass could not be manufactured in the colonies. Braziers were completely dependent on England for their materials. Large quantities of brass lighting devices were imported to the colonies from England, particularly from the manufacturers in Birmingham. Braziers could remake objects from broken or worn brass pieces. Brass was primarily formed by sand casting, so braziers were also pattern makers and served trades such as pewtersmithing, which required brass molds. Forms cast by braziers included candlesticks, chandeliers, sconces, gun parts, fireplace tools, and bells. Because casting is a hot and dirty process, the career of the brazier was one of the less desirable metal trades.

PEWTER

Pewterers, like braziers and silversmiths, were dependent on England for raw materials and spent much of their time repairing or replacing imported British goods. Most pewterers also sold new, imported pewter wares from England. Domestic pewter was imported more than any other metal goods.

Pewter was usually formed by casting it into brass molds. Immigrant pewterers brought their molds with them from England. These were expensive to replace and so styles of pewter were slower to change than those of copper or silver. Symmetrical objects like cups or bowls would then be put on a **lathe** and scraped of excess material and burnished to a high polish. This also meant that pewterers required an assistant to turn the lathe, unlike other metal trades that could be operated singly. The first pewter shop in the colonies was opened in 1635. The first pewter forms made in the New World were probably spoons and buttons.

Pewter was a universal material on the tables of the middle and upper classes. Only the wealthiest used silver. Common pewter goods included **flatware,** plates, platters, cups, and bowls. When buying pewter, the purchaser often had his name or initials stamped or engraved on them to deter theft. In England, the Worshipful Company of Pewterers closely oversaw the standards of pewter quality and workmanship. This was impetus for colonists to purchase imported goods rather than have wares made locally, where there was no such assurance.

IRON

Iron was the exception to the colonists' general reliance on materials from England. At the time, England was growing dependent on Sweden for iron and sought another source. Colonists were smelting iron ore by 1648. After smelting, the iron was sand cast into the desired shapes.

One popular product of iron workers was the fireback. This was a large plate of cast iron placed at the back of a fireplace to protect the bricks. These were ornamented with decorative motifs of the period. Other cast iron products included stoves, andirons, kettles, cooking pots, laundry iron, tongs, guns, grave markers, hinges, and other architectural hardware.

See also Baroque; Connections; First Metalsmiths; Nineteenth Century; Rococo and Neoclassical; Tiffany & Company; Yellin, Samuel

FURTHER READING

Mitchel, James R., ed. *Antique Metalware: Brass, Bronze, Copper, Tin, Wrought and Cast Iron.* New York: Universe Books, 1976.

Colonial Williamsburg, Williamsburg, Virginia (includes The DeWitt Wallace Decorative Arts Museum), http://www.history.org

Fales, Martha Gandy. *Early American Silver,* revised and enlarged edition. New York: E. P. Dutton & Co., 1973.

Hopewell Furnace, National Historic Site, Pennsylvania, http://www.nps.gov/hofu

Kauffman, Henry J. *American Copper and Brass.* Camden, NJ: Thomas Nelson, 1968.

Kauffman, Henry J. *The American Pewterer: His Techniques and His Products.* Camden, NJ: Thomas Nelson, 1970.

Kauffman, Henry J. *The Colonial Silversmith: His Techniques and His Products.* Camden, NJ: Thomas Nelson, 1969.

Montgomery, Charles F. *A History of American Pewter,* revised edition. New York: Dutton, 1978.

Shelburne Museum, Shelburne, Vermont, http://www.shelburnemuseum.org

Winterthur Museum and Country Estate, Winterthur, Delaware, http://www.winterthur.org

Connections

Metals may be connected in many ways. These may be divided into hot and cold, depending on whether or not the parent metals being joined are heated during the process. The methods listed below are meant to be demonstrative but not exhaustive.

HOT CONNECTIONS

Hot connections are usually more secure than cold ones and are only used between two or more metals.

Brazing or Hard Soldering

Brazing and hard soldering are two terms used for the same joining process, in which a third metal is introduced in order to join two parent metals at a temperature greater than 450°C (842°F), but lower than the melting temperature of the parent metals. This technique may be used when the metals to be joined are either the same or different from one another, such as silver to silver, or silver to copper. The third metal added as a bonding agent is usually an alloy (see **Metals**). As the parent metals approach their melting point(s), the crystalline structure of their surfaces expands, and the alloy melts and diffuses into the irregularities created. On a crystalline level, this added metal intermingles with both the parent metals and thus joins them together.

When the alloy being used as a joining material is primarily made of silver, this is sometimes called silver soldering. Unless otherwise specified, brazing generally implies the use of a brass alloy as the filler. Hard soldering/brazing is the joining process generally used by silversmiths, goldsmiths, coppersmiths, and braziers.

Soft Soldering

Soft soldering is very similar to hard soldering. The biggest difference is that soft soldering occurs below 450°C (842°F), usually around the 300°C–400°C range. A metal with a lower melting temperature than the two parent metals is added while heating all parts. Because the parent metals are not approaching their melting point(s), there is no intermingling of the added alloy and the parent metals on a crystalline level. This results in a weak joint both because of the superficiality of the bond and the weakness of the filler metals used. The filler metal in this case is generally an alloy of tin, lead, and sometimes bismuth. Soft soldering is commonly used for copper plumbing and was first implemented by the ancient Romans.

Welding

Welding is a means by which similar or identical metals are joined by melting them directly together. Unlike brazing or soldering, a welded joint is as strong as or stronger than the parent metals. The molten state of the parent metals is achieved through the application of heat or electricity. There is evidence of welding being used in nonferrous metals as early as the third millennium B.C.

Forge Welding

In forge welding, the two metal parts to be joined are brought very close to their melting temperature, lain one atop the other, and struck with a hammer. The heat and force cause the intermingling of the crystalline structures of the two parent metals. Early dynastic Egyptians used forge welding to make long, continuous pieces of copper wire.

Gas Welding

Gas welding involves the use of a flammable gas torch to melt metal together. A common setup of this kind involves the pairing of acetylene, a flammable gas, with oxygen, which helps the acetylene burn hotter and cleaner. This is commonly called oxyacetylene welding. The areas to be joined on the parent metals are placed in contact and brought to melting temperature as a rod of similar material is also melted into the seam. This filler rod strengthens the seam by providing additional material. Thus, the molten metal flows together and then cools into a continuous joint. This kind of welding has been in use since the end of the nineteenth century.

Electric Arc Welding

In electric arc welding, or simply arc welding, heat from an electrical arc is used to melt metal together. The parent metals are attached to one side of an electric circuit and a metal electrode of the same or similar material is placed on the other side of the same circuit. When brought into close proximity, the current causes a continuous spark across the gap called an electric arc. The heat produced by the arc melts the parent metals together. The electrode itself also melts to act as a filler rod. The electric arc has been used to join metals since the late nineteenth century.

MIG Welding

MIG stands for "metallic inert gas." As in arc welding, the heat for welding is produced by an electric arc. Instead of a filler rod, however, a hand piece connected to the electrical source feeds a continuous wire of the same metal that is being welded, thus acting both as the electrode and the filler material. The hand piece also emits an inert gas that protects the molten metal from oxidation while forming the weld, thus making for cleaner welds. MIG welding came into use around 1948.

TIG Welding

TIG stands for "tungsten inert gas." In this process the heat for welding is produced though the creation of an electric arc between a tungsten rod and the parent metals at the other side of the electric circuit. An inert gas is released from the same hand piece where the tungsten rod is located, similar to MIG welding. The presence of an inert gas shield also makes it much easier to weld difficult metals such as aluminum. A handheld filler rod may be used as well. This method of welding was developed in 1941.

COLD CONNECTIONS

Cold connections are generally weaker and more easily reversed than hot ones. These may also be used to connect metal to other materials like stone, wood, or plastic, which would be damaged by heat.

Rivet

A rivet is a rod or wire that passes through holes in two or more materials and becomes thicker on the end or ends where it protrudes in order that it may not be removed. There is evidence of rivets being used as early as 3000 B.C.

Tab

A tab is a small extension from one piece of metal that bends around an object or other piece of metal in order to hold the two together.

Collar

A collar is a band of metal that is fashioned around two or more objects to bind them together. Blacksmiths generally use collaring to join parallel, linear elements.

See also Forging; Gilding and Plating; Inlay; Lamination Techniques

FURTHER READING

Andrews, Jack. *New Edge of the Anvil: A Resource Book for the Blacksmith.* Drexel Hill, PA: SkipJack Press, 1994.
The Artist-Blacksmith's Association of North America, http://www.abana.org
Briggs, Charles W., Ivan H. Griffin, and Edward M. Roden. *Basic Welding Techniques.* New York: Litton Educational Publishing, 1977.
McCreight, Tim. *The Complete Metalsmith,* professional edition. Portland, ME: Brynmorgen Press, 2004.
Untracht, Oppi. *Metal Techniques for Craftsmen.* Garden City, NY: Doubleday, 1968.
Welding Information Center, http://www.weldinginfocenter.org

Copper. ***See*** Metals

de Lamerie, Paul
(1688–1751)

Paul de Lamerie was one of the earliest and most well-known silversmiths working in the English **rococo** style. His workshop was extremely prolific and responsible for some of the finest silver pieces of the period. His peers recognized him as an outstanding craftsman, and his career exemplifies the English silver trade of the early seventeenth century.

De Lamerie's parents were minor nobility and Huguenots, French Protestants, who had left France under the threat of religious persecution. They spent a short time in Holland, where Paul was born, before finally settling in England. When Paul was 15 he was apprenticed to Peter Plattell, a goldsmith in London who was also a Huguenot of minor nobility (see **Guilds**). Paul completed his apprenticeship in 1711 and stayed on in Plattell's shop while he saved enough money to open his own business. On February 4, 1713, de Lamerie registered at

THE HUGUENOTS

On April 13, 1598, King Henry IV of France issued an edict protecting the rights of the Protestant minority after 50 years of religious division and warfare between the French Catholics and the Protestants. The French Protestants, or Huguenots, lost their freedom of worship when the Edict of Nantes, as the edict was known, was revoked by Louis XIV in 1685. Following this, around 200,000 Huguenots fled to other parts of Europe or North America. Many were skilled craftsmen, and they brought prosperity and revolutionary French technologies to the places in which they settled. Significantly, many French silversmiths had converted to Protestantism in the sixteenth and seventeenth centuries. The sudden influx of these talented and interconnected craftsmen had a lasting impact on the silver industries in England and North America.

the Goldsmith's Hall of London and became a Freeman with all the privileges of a full citizen. In that same year de Lamerie set up his first of many shops.

Initially, de Lamerie worked in the two styles that were most fashionable at the time: the simplified Queen Anne style and the more ornamented Huguenot style that included repeated decorative elements and more intricately **cast** parts. It is not until the appearance of a fruit dish from his shop in 1734 that we see the first evidence of the rococo style for which Paul de Lamerie would be known.

Originating in France and used extensively in the eighteenth century, the rococo style incorporated elaborate ornamentation in naturalistic motifs, especially leaves, shells, and water-worn rock formations. It was characterized by asymmetry and the use of scrolls in the forms of the letters "c" and "s." The rococo style was used in all the major decorative arts, although silver was the pioneering medium for the style. Such undulating surfaces caused a great play of light when rendered in polished gold or silver. Inventiveness is another important aspect of the rococo, and de Lamerie's designs were known for both their technical excellence and design innovation.

The esteem of his peers was made most evident in 1740 when Paul de Lamerie was commissioned to create an ewer and basin for the Goldsmith's Company. They ordered the pieces for ceremonial display, to be created in silver and plated with gold. These were executed while de Lamerie was at the pinnacle of his mature rococo style. They are a tour de force of all the silversmith's techniques that were in use in that period.

The ewer is shaped like an upside down helmet mounted on a pedestal foot, which was a characteristic form used by many Huguenot silversmiths. The body of the vessel is decorated lavishly with fantastic imagery from the sea including winged mermaids, sea monsters, and a handle that terminates in a relatively large triton figure. The cast arms of the Goldsmith's Company are applied below the vessel's pouring spout.

The basin's center also contains the Goldsmith's Company's coat of arms, which were cast and applied. The surrounding border is heavily modeled in relief with contrasting textures that were achieved through **chasing and repoussé, engraving,** and casting. The four oval cartouches, or framed areas, around the ewer's edge contain figurative scenes. These feature *putti,* small flying babies, dressed as characters from Roman mythology: Vulcan, Mercury, Hercules, and Minerva. Between them are animal representations of the four elements: lion (earth), sea monster (water), phoenix (fire), and falcon (air).

Although these magnificent pieces bear Paul de Lamerie's maker's mark, they were most likely the work of many artisans organized by him. There is a tendency to imagine that a maker's mark indicates absolute authorship of the work that it appears on. However, it is literally impossible that de Lamerie could have made every piece that was marked with his insignia. It is most likely that in addition to those who he employed in his workshop, de Lamerie also made use of an extensive network of independent craftsmen. His place within the strongly interconnected Huguenot community would have aided significantly in that respect.

There has been evidence of specialization in the trade of silversmithing from as early as the fifteenth century. It was not at all unusual in the eighteenth century for a piece to pass through the hands of an individual designer, a plate worker, several specialized ornamenters, and a goldsmith-retailer. Only the latter would have his mark on the finished object.

Thus, some objects that bear de Lamerie's maker's mark may not have been worked on in his shop at all. It was not uncommon for a registered goldsmith to buy finished pieces from other goldsmiths and sell them under his own mark as the retailer, especially if the original maker was not registered. These shrewd business practices enabled de Lamerie to provide a wide array of objects to the market.

See also Baroque; Plating and Gilding

FURTHER READING

Alcorn, Ellenor M. *Beyond the Maker's Mark: Paul de Lamerie Silver in the Cahn Collection.* Cambridge, UK: John Adamson, 2006.

Carver, Beth S., and Eileen M. Casey. *Silver by Paul de Lamerie at the Sterling and Francine Clark Art Institute.* Williamstown, MA: The Institute, 1978.

Hare, Susan, ed. *Paul de Lamerie: At the Sign of the Golden Ball: An Exhibition of the Work of England's Master Silversmith.* London: Goldsmith's Company, 1990.

Phillips, Philip Alexander Solomon. *Paul de Lamerie, Citizen and Goldsmith of London, a Study of His Life and Work, A.D. 1688–1751.* London: B.T. Batsford, 1935.

Design Reform. ***See*** Arts and Crafts Movement

Electroforming

Using electrolysis to deposit a layer of metal on an object is referred to as electrodeposition. In electroforming, the layer created is thick enough to exist on its own, thus constituting a method by which a metal object may be formed. The positive end of an electrical circuit, or anode, is hooked up to a piece of the metal being deposited. The object to be plated acts as the negative electrode, or cathode. The transference takes place in various liquids known as an electrolytic solution. During electrodeposition, ions of metal pass through the electrolyte from the anode to the cathode, where they build up over time. The support that the metal is deposited on must be capable of conducting electricity. Nonmetallic substrates are usually coated with a liquid in which very small pieces of metal are suspended in order to make them conductive. Electrotyping is the use of electrodeposition over a mold or model of an object in order that the object may be duplicated. In such cases, the original is made of wax and frequently gets melted out after the electrodeposition has taken place. The first successful electrodeposition of metal was performed some time around the year 1800.

See also Gilding and Plating; Metals; Schwarcz, June

FURTHER READING

Curtis, Leslie. *Electroforming*. London: A & C Black, 2004.
Electrochemical Science and Technology Information Resource, http://electrochem.cwru.edu/estir

Electrum. ***See*** Metals

Embossing. ***See*** Chasing and Repoussé; Stamping and Die Forming

Enamel

Enamel is powdered glass that may be colored with metallic oxides. Enamel is fused to the surface of certain metals in order to obtain a glasslike finish. The term enamel also refers to the resulting object, once it has been enameled. The best metals on which to enamel have compatible expansion and contraction rates when raised to the fusing temperature of the enamel. This is usually around

800°C or 1,450°F, which is achieved through the use of a kiln. Traditional metals for enameling include silver, gold, and copper. Other metals that have been enameled include bronze, iron, and steel. Enamel may be used as a powder that is sifted, in a paintlike liquid suspension, or as lumps or threads (thin rods). Enamel may be colorless, opaque, transparent, or opalescent in color.

CLOISONNÉ

The cloisonné technique of enameling involves the formation of small cells on top of the surface of the parent metal. These cells are created with strips of wire in gold, silver, copper, or brass that are bent to shape and then filled with enamel. After the cells are filled, the design is abraded until the surface is smooth and the tops of the wires are flush with the enamel. The term cloisonné comes from the French term for cells, called *cloisons*. The use of cloisonné originated as a means of mounting precious stones into place with cement, resulting in something like a mosaic.

CHAMPLEVÉ

In the champlevé technique, recessed areas to be filled with enamel are created within the body of the parent metal. This recess may be cast in place or created through etching, chiseling, engraving, or fabrication. Frequently, large planes of unadorned metal contrast the enameled recesses. The term *champlevé* comes from a French phrase meaning "raised field."

BASSE TAILLE

Basse taille means "shallow cut" in French, but when it was first developed in thirteenth-century Italy it was called *lavoro di basso rilievo* (low relief work). Basse taille is an enameling technique characterized by the use of transparent enamel over metal with a decorative texture. This is usually done on silver or gold in order to better reflect light through the enamel. The texture underneath was traditionally engraved or chased, although contemporarily it might be achieved in many more ways. In the *guilloche* technique, an engine-turned **lathe** cuts a repetitive, linear pattern that may be covered in transparent enamels.

PLIQUE-À-JOUR

One of the most technically complex forms of enameling, plique-à-jour enamel spans open spaces without a backing to support the fused glass. The enameling is done on an open metal framework with a removable back, such as a thin piece of metal or mica. The enamel used is usually translucent or transparent, resulting in an effect similar to a stained glass window. *Plique-à-jour,* meaning "light of day," was developed in fourteenth-century France.

See also Ancient Rome; Byzantine Empire; China; Fabergé, Peter Carl; Islamic Metalwork; Middle Ages; Renaissance; Schwarcz, June

FURTHER READING

Benjamin, Susan. *Enamels: The Smithsonian Illustrated Library of Antiques.* New York: Cooper-Hewitt Museum, 1983.
Darty, Linda. *The Art of Enameling.* New York: Lark Books, 2004.
The Enamelist Society, http://enamelistsociety.org
Glass on Metal (periodical). Newport, KY: Ceramic Coating Co., 1982–present.
Jazzar, Bernard, and Harold B. Nelson. *Painting with Fire: Master of Enameling in America, 1930–1980.* Long Beach, CA: Long Beach Museum of Art, 2007.
Matthews, Glenneth Lesley. *Enamels, Enameling, Enamelists.* Iola, WI: Krause Publications, 1984.

Engraving

Engraving is the process by which lines are incised into a metal surface to produce a decorative finish. The tool traditionally used is a small, pointed chisel that simultaneously cuts and lifts the metal to remove it, called a graver or a burin. The short, round handle of the tool is held low in the palm, while the cutting end is held between the thumb and forefinger. The other fingers help to hold the handle and control the steady, forward cutting motion. The technique of engraving metal dates back to around 1000 B.C. Although the act of scratching designs into soft materials goes back much further, true metal engraving was not possible until sharpened steel implements could be produced. Most engraving performed contemporarily is done with power tools such as pneumatic gravers or fine rotary grinders. Engraving may be used in conjunction with **enamels** and **niello.**

In bright-cut engraving, short, broad pieces of metal are gouged out to produce incised facets meant to sparkle in the light. This kind of engraving came into vogue with silver engravers in the second half of the eighteenth century. In wrigglework, an engraving tool with a small flat area at its cutting edge is walked across the metal surface with a rocking motion that produces a zigzag line instead of the traditional smooth one. Wrigglework was a popular way to depict imagery on pewter and silver in the Netherlands and England from the seventeenth and eighteenth centuries.

See also Etching; Flatware; Inlay

FURTHER READING

Alfano, Sam. "The Engraver's Café," http://igraver.com
Hasluck, Paul N. *Engraving Metals.* Bradley, IL: Lindsay Publications, 2001.
Oman, Charles. *English Engraved Silver, 1150–1900.* London: Faber & Faber, 1978.

Etching

Etching is a decorative technique used to selectively remove metal through controlled corrosion, thereby creating patterns of contrasting heights. The two main components of etching are a resist and a mordant. The resist protects the areas of metal that are to remain intact. The mordant is the substance, usually an acid, that eats away at the metal to remove it. The kind of metal being etched determines what mordant is used. The resist is chosen in accordance with the mordant as well as the requirements of the intended design. Solutions of nitric acid or hydrochloric acid have been used extensively as mordants. Today many metalsmiths use ferric chloride, which is a strong salt, as a safer way to etch copper-based alloys. Some examples of resists include lacquer, varnish, pitch, wax, bitumen, and resin.

Examples of etching can be found as far back as the late thirteenth century A.D. By the fifteenth century it was being used extensively to add surface decoration to armor. In the following century, etching was adopted for use on small domestic objects in nonferrous metals and was also being implemented as a printmaking technique.

See also Armor and Arms; Enameling; Engraving; Niello; Schwarcz, June

FURTHER READING

Untracht, Oppi. *Jewelry Concepts and Technology*. New York: Doubleday, 1982.

Walker, Perrin. *CRC Handbook of Etchants for Metals and Metallic Compounds*. Boca Raton, FL: CRC Press, 1990.

Fabergé, Peter Carl
(1846–1920)

Peter Carl Fabergé was the best-known member of a family of goldsmiths in Russia at the end of imperial rule. The Fabergé name was synonymous with luxurious articles of surpassing technical excellence. He is now most famous for designing a series of elaborately fabricated eggs, given as gifts within the imperial household at Easter.

Peter Carl Fabergé's father, Gustav, was a goldsmith who was most likely descended from Huguenots (see **Paul de Lamerie**). Gustav operated a shop in St. Petersburg, Russia, from 1842 to 1860. Peter Carl Fabergé, usually called Carl, was born in May of 1846. His father retired to Dresden in 1860, leaving managers in charge of the shop. Carl had been brought up in the family business and was sent on a tour of Germany, Italy, and France to complete his goldsmithing education. He returned to the family business in St. Petersburg around 1866 and by 1872 had assumed control.

At first the work executed at the shop was similar to many other Russian jewelers and not very distinctive. Seeking to ingratiate himself with the nobility, Carl performed free repairs and appraisals for the Hermitage Museum. This may have led to the invitation to participate in the Moscow Pan-Russian Exhibition of 1882. With this exposure, attention and commissions from the royal court followed.

From 1882 until his death in 1895 Carl's brother, Agathon, was also a part of the family business. The greatest stylistic advances were made in the firm during the period in which the brothers worked together. It was at this time that the most fantastic and imaginative devices were created. Some speculate that these creative endeavors were the result of Agathon's exposure to work by goldsmith Johan Dinglinger while in Dresden. Other sources of inspiration may have come from Carl's extensive travels or the nearby Hermitage Museum. The style of Fabergé was largely an amalgam of revival styles, including Louis XV, Louis VXI, and **Renaissance.** It also, at times, incorporated influences from **art nouveau, Japan,** Russian history, and the empire style.

A TRADITION OF COURTLY AMUSEMENTS

Prior to the age of mass media, the products of a court goldsmith were expected to play the part of conversation pieces for the nobility. Entertainment was derived from clever devices or intricate reproductions of fantastic scenes. One celebrated example of this function is the Court of the Grand Mogul by Johann Melchior Dinglinger (1664–1731). Dinglinger was the court goldsmith of August the Strong in Dresden, Saxony. The Court of the Grand Mogul is a cabinet piece (object for display) executed in the late **baroque** style. It is a tableau of the exotic Indian court as the Grand Mogul Aureng Zeb receives his birthday presents. Executed in gold, silver, precious gems, and enamel, the stagelike display is a feat of technical brilliance. Measuring approximately four feet wide and deep, the scene includes 132 figures, each two and three inches high, as well as several exotic animals. The combination of richness and exoticism presented by such a representation was a diversion fit for a king.

In 1885 Czar Alexander III commissioned an Easter egg for his Empress Maria Feodorovna on the occasion of their twentieth anniversary. The result was the first of the famed Fabergé imperial Easter eggs. The first egg, known as the Hen Egg, appeared to be a simple white egg, executed in **enamel.** When opened, a golden yolk was revealed, and within the yolk was a golden hen. When the hen was opened a miniature crown and a tiny egg-shaped jewel were revealed. The gift was so well received that the czar ordered that one be made each year. The only constraints were that they all had to be egg-shaped, no two be alike, and that there should always be a surprise inside. Thus in 1885 Fabergé was awarded the title of Supplier to the Court of His Imperial Majesty.

In future years, the eggs became more splendid and elaborate. When Czar Alexander III passed away, his son, Czar Nicholas II continued the tradition by presenting one egg annually to both his empress and his mother. Some eggs were clocks, while others contained delicate miniature paintings of family members or the places they loved. The egg given by Nicholas II to his wife, the Empress Alexandra Feodorovna in 1897, called the Coronation Egg, contained a perfect working miniature of the coach in which the empress had ridden to the coronation ceremony. Some eggs were transparent, revealing their surprises through walls of rock crystal. Most of the eggs took a year or more to make.

The increased stature that followed the royal commissions brought a boom in business from the nobility and rich merchants. As the Fabergé business grew, it became necessary to employ a number of head work masters to oversee the execution of objects by specialists. Operations expanded, and a branch was opened in Moscow in 1887. By the turn of the century, Fabergé was the most important undertaking of its kind in Russia. More branches were eventually opened in Odessa, Kiev, and London. At the peak of prosperity, Fabergé employed some 500 people, including designers, modelers, stone carvers, enamellers, and goldsmiths. Carl only designed the most important objects, such as the imperial Easter eggs. His sons Eugène, Agathon, and Alexander joined the family business as well.

In addition to the Easter eggs, other objects of fantasy created by the House of Fabergé included facsimiles of flowers in vases, executed in carved stones, gems, and precious metals. Carved stone miniatures of people and animals inspired by Japanese netsuke were very popular and widely purchased from Fabergé. Other kinds of objects produced by the firm included silver dinner services and centerpieces, cigarette cases, small picture frames, and letter openers. High standards of technical excellence were always maintained, with an emphasis on very limited or one-of-a-kind production. The House of Fabergé appears to have excelled particularly in *guilloche* enameling and lapidary work (gemstone carving).

The Fabergé fame was spread abroad through exhibitions and by the travels of the royal family. When the czar's family went abroad, they brought with them rich gifts from the workshops of Fabergé. Articles from the House of Fabergé were also given as gifts at royal weddings, coronations, and betrothals. The Fabergé name was synonymous with the elegance of the lives of the Russian nobility.

The extreme luxury of the nobility was in sharp contrast to the lives of the peasantry at the time. Russia's participation in World War I was followed by revolutions, the abdication of Czar Nicholas II, and the eventual execution of the royal family. Carl Fabergé escaped to Lausanne, Switzerland, where he died in 1920.

See also Baroque; Guilds; Nineteenth Century; Rococo and Neoclassical

FURTHER READING

Bainbridge, Henry Charles. *Peter Carl Fabergé, Goldsmith and Jeweller to the Russian Imperial Court: His Life and Work.* London: Spring Books, 1966.
Fabergé Arts Foundation, Washington, DC, http://www.fabergearts.org
Fabergé, Tatiana, Lynette G. Proler, and Valentin V. Skurlov. *The Fabergé Imperial Easter Eggs.* London: Christies, 1997.
Habsburg, Géza von, and M. N. Lopato. *Fabergé: Imperial Jeweler.* St. Petersburg, Russia: State Hermitage Museum, 1994.
PBS Treasures of the World, http://www.pbs.org/treasuresoftheworld/index.html
The State Hermitage Museum, St. Petersburg, Russia, http://www.hermitagemuseum.org

Filigree

Filigree is an ancient technique of building up decorative forms through the use of wires. Delicate tendrils of wire are woven, wrapped, twisted, and affixed into generally planar formations, frequently resembling lace or twining botanical elements. Filigree may be made with or without a backing material. Backing materials are generally either sheet metal or prefabricated wire mesh. The filigree is attached to its backing with either solder or cold **connections.** In unbacked filigree, a framework of heavier wires must be used to give the smaller wires stability and structure. The larger pieces are always fabricated first in order that successive soldering operations may occur at lower temperatures. The solder used to connect the smallest wire elements is dusted on the piece as a fine powder. The effect of filigree wire can also be used in conjunction with a planar filling material, such as **niello** or **enamel.** With enamel, the filigree wires work much like cloisonné wires. Greater variety may be achieved by changing the shape of the cross section of the wires used.

Traditionally, filigree was executed in gold and silver but from the nineteenth century on has been more frequently made in base metal alloys and then **plated** for large-scale production. Filigree has been used as far back as **ancient Egypt.** Although it is more frequently used in jewelry, it has also been incorporated to great effect in sword fittings, horse trappings, domestic vessels, medieval reliquaries, book covers, **flatware,** and even crowns. In addition to being both beautiful and visually complex, well-executed filigree is also lightweight and stable. Furthermore, it allows a small amount of material to make a big visual impact. The cost of materials is relatively small, but the work tends to be time consuming. This technique continues to be produced most in areas where the cost of labor is relatively low.

See also Byzantine Empire; Judaica; Middle Ages; Nineteenth Century

FURTHER READING

Menshikova, Maria, and Jet Pijzel-Dommisse. *Silver Wonders from the East: Filigree of the Tsars.* Aldershot, UK: Lund Humphries, 2006.

Untracht, Oppi. *Metal Techniques for Craftsmen; A Basic Manual for Craftsmen on the Methods of Forming and Decorating Metals.* Garden City, NY: Doubleday, 1968.

First Metalsmiths

As the Stone Age drew to a close, some of the formerly nomadic peoples began to raise crops and domesticate animals, creating agrarian villages. This major development in the history of civilization allowed for greater stability and the exploration of new technologies.

Thus, around 6500–5500 B.C. the chalcolithic age began in some parts of the world. The chalcolithic age was characterized by the settling of villages and the use of copper for the production of some tools, adding to the already pervasive stone ones. Its inception began in western Asia near the Tigris and Euphrates Rivers, followed by similar developments in the Nile River Valley and the areas surrounding the Aegean Sea, and later by Europe, **Africa,** and eventually eastern Asia. As subsequent metalworking technologies were developed, they generally followed a similar route of diffusion out from western Asia and the eastern Mediterranean region. The spread of the new agrarian lifestyle necessarily preceded the working of metals. Most of the ancient world was using metals of some kind by 2000 B.C.

Metalworking technologies dispersed slowly and were not immediately adopted when introduced. The sequence in which the different metals were exploited was generally the same, beginning with gold and copper, followed by bronze, silver, nonmeteoric iron, and finally steel (see **Metals**). The geological characteristics of a region sometimes caused a step in this progression to be skipped. For example, there was no period of bronze use prior to iron in Japan, and no copper period at all in most of Britain.

Metals that occur in a pure form in nature, such as gold, copper, meteoric iron, and electrum, were the first discovered and used by man. As early as 5000 B.C., people were picking up copper and gold and trading them for ornamental use. It was discovered that copper could be formed through force and in the process became more rigid. Once melting and casting were developed, copper was used for tool making, and the softer gold remained ornamental. Gold was used for personal adornment as far back as 4000 B.C. Early examples contain traces of other metals, frequently silver, as purification methods were not developed until the first millennium B.C.

Pure copper from nature is a limited resource, and to use copper in any quantity it is necessary to smelt it from ore (see **Mining and Smelting**). The principle of smelting was probably discovered around 4000 B.C. through the process of firing ceramics at high temperatures. Such heat would reduce malachite, a seemingly ordinary stone, into the malleable metal copper. Metals were smelted in the hills where ores were mined and trees for fuel were found. The practice of actively mining metals began as early as the fourth millennium B.C. in some regions and probably developed out of the digging for flint.

The discovery of bronze was most likely the result of smelting ores that had both copper and tin content. Conscious alloying was being done in Egypt and Mesopotamia by 2500 B.C. Tin does not occur in a pure form in nature, and tin-bearing ores are much less widespread than copper-bearing ores. Bronze production was initially limited to areas where both ores could be found. The resulting metal was then traded widely by itinerant bronze-smiths.

The great advantages of bronze over copper are its hardness and ability to be cast much more easily. Because areas of Europe had greater access to tin, central Europe led the early world's production of bronze by 1500 B.C. To speak of a "Bronze Age," a period characterized by use and technical mastery of that metal, would include vastly different time periods for every region.

Lead, which is found in the mineral galena, was probably discovered by accident in a campfire due to lead's low melting temperature. When the ore undergoes cupellation, heating in a container of bone ash, the byproduct is a small amount of pure silver. This process was being used widely around the Mediterranean by 2500 B.C. in order to render silver.

The earliest evidence of iron use dates from the third millennium B.C. Meteorites sometimes contain pure, workable iron, but they are rare compared to the iron ores available in the earth's crust. Iron smelting grew out of copper smelting but required a higher heat and a low-oxygen atmosphere. The first smelting furnaces were probably charcoal pits fitted with bellows that made the fire burn hotter. Smelting iron results in a chunk of material called a *bloom.* The bloom must be heated until it glows red and **forged** many times in order to drive out the nonmetallic elements and produce a useable material. This useful iron was not made in quantity until around 2000 B.C., starting an Iron Age for some.

Early metalsmiths did not have the technology to create a heat great enough to melt and cast iron, so it had to be formed through forging. Iron is only able to replace bronze for impact tools when it is transformed into steel. This requires the addition of carbon and the knowledge of hardening and tempering methods. These technologies were first seen in the second millennium B.C.

The first metalworkers shaped their material by hammering and grinding it with rocks. They used the sinking method to form metal into recesses in stone or wood (see **Raising**). By 5000 B.C. metalworkers were **annealing** metal to improve its malleability. Initially open stone molds were used for **casting,** but by the third millennium B.C., the lost wax method was being used in some locations. As the practice of metalworking continued, specializations developed. In early cities, one could find work that had been riveted, **engraved,** soldered, **inlaid,** spun, and welded (see **Lathe Work; Connections**).

The first metal forms copied existing objects made in stone and bone until the advent of casting. Natural copper was used to make small things like beads, ribbon, tweezers, awls, needles, and fishhooks. After smelted copper was available, larger tools like chisels, axe heads, weapons, harpoon heads, vessels, helmets, shields, and saws and chisels for wood were made. Chalcolithic weapons were limited to short cutting implements like daggers. It was not until the development of bronze that longer, more saberlike weapons developed. Bronze was also used for large statues and architectural elements. When the Iron Age began, steel replaced bronze as the desirable metal for tools and weaponry. In the chalcolithic period, lead was sometimes used for vessels and figurines. Gold and copper were also beaten into thin sheets to clad sculpted wooden armatures (see **Gilding and Plating**).

Objects made of metal were primarily used in the service of religious observation or as personal possessions for the wealthy and powerful. In the two major pre-Hellenistic societies, Egypt and Sumer, metalworkers were placed apart for their seemingly magical abilities to draw forth metal from ores and work them into forms. At every point in the ancient world metals were rare and costly and were frequently traded from great distances.

See also Africa; Ancient Greece; Ancient Rome; Armor and Arms; Celtic People; China; India; Islamic Metalwork; Japan; Persia; Pre-Columbian

FURTHER READING

Aitchison, Leslie. *A History of Metals.* New York: Interscience Publishers, 1960.

Haedeke, Hanns-Ulrich. *Metalwork,* Vivienne Menkes, trans. New York: Universe Books, 1970.

Huygh, René, ed. *Larousse Encyclopedia of Prehistoric and Ancient Art; Art and Mankind.* New York: Promethius, 1962.

Knauth, Percy. *The Metalsmiths.* New York: Time-Like Books, 1974.

The Metropolitan Museum of Art, New York, http://www.metmuseum.org

The State Hermitage Museum, Prehistoric Art Collections, St. Petersburg, Russia, http://www.hermitagemuseum.org

Flatware

Flatware is the collective name for the table utensils that are used to serve and consume food. These are also sometimes referred to as cutlery. The most common examples of flatware are the knife, spoon, and fork. Although today these utensils are generally purchased in matching sets, each one has a unique history as an autonomous object.

Early implements were made of organic materials such as wood, bone, and shell. As metalworking technologies developed, however, they were made in bronze, iron, and steel. Eventually, the wealthier classes had flatware of silver, gold, or gold plate. The handles of these expensive utensils might have been made out of any number of materials, including silver, ivory, or **enamel.** Until the nineteenth century, the general populous had utensils of wood, iron, steel, copper-based alloys, or pewter. From the nineteenth century onward, nickel alloys and electroplated wares mimicked the appearance of silver (see **Gilding and Plating**). The development of stainless steel in the second decade of the twentieth century revolutionized the flatware industry. Being both hygienic and durable, this was the perfect choice for households without servants to care for silver. It also accommodates the increasingly relaxed dining style of most homes and now makes up all but five percent of flatware production.

Prior to the industrial revolution, flatware was produced by hand in workshops. Historically, the methods of manufacturing flatware have included **forging,** cutting, and **casting.** The invention of steam-powered die stamping allowed for faster manufacture, although it somewhat limited ornamentation (see **Stamping and Die Forming**).

The first examples of matching flatware were dessert fork and spoon sets from the seventeenth century. At the time, dessert was considered a separate meal. It was also not until the seventeenth century that hosts began to provide flatware for their guests. Matching sets that included a knife, fork, and spoon for general dining developed in the late eighteenth century.

THE KNIFE

The knife is the oldest and most pervasive of all eating utensils and originally doubled as a weapon. In the Middle Ages knives were used to cut and serve meat. At that time and for centuries afterward, only the wealthiest members of society had their own knives to eat with, which they carried with them. The less wealthy shared communal knives at the table.

Knives meant solely for use at the dinner table did not really exist before the fifteenth century. In France a law was passed in 1669 forbidding the use of pointed table knives, which led to their being manufactured with a rounded end. The flat expanse of that blade was used to convey food such as peas to the mouth.

ICE CREAM KNIFE?

Toward the end of the nineteenth century in the United States silver flatware was more accessible than it had ever been before. The price of silver was driven down when large deposits like the Comstock Lode in Nevada were found. Electroplating made silver-plated flatware an even less expensive option. Manufacturers took advantage of this availability to create new decorative patterns and specialized flatware forms that had previously never existed. The invention of things like the griddlecake server, berry scoop, and ice cream fork had some American silver sets containing up to 146 pieces. By the 1920s the increase in specialized flatware was so exorbitant that the federal government felt the need to intervene. Herbert Hoover, who was secretary of commerce at the time, decreed that silver services be limited to 55 pieces each.

THE SPOON

The spoon was developed as a means to deal with liquid. It is from **ancient Egypt** that the earliest metal spoons survive. From antiquity there is evidence of spoons made in bronze silver, and gold. They initially had cylindrical handles. The flatter handles that are more common now were developed in the seventeenth century

The use of silver spoons was well spread throughout the European aristocracy by the seventeenth century. In the second half of the seventeenth century, a rising middle class in many countries began to purchase silver tableware. A silver spoon was frequently the first article of silver that a family would purchase. Commemorative silver spoons were given to children at births and christenings, to women when they wed, and to mourners at funerals. In the nineteenth and twentieth centuries the souvenir spoon became a standard collectible and could be found at expositions, worlds fairs, and tourist destinations.

THE FORK

Although there is evidence of the fork's use in antiquity, it was primarily a serving tool. Personal forks were used in the **Byzantine Empire** as early as the sixth century. Individual forks for eating sweets called sucket forks appeared in Western Europe by the eleventh century. Their use in conjunction with the knife for meat eating began in Italy in the sixteenth century and spread slowly throughout Europe.

The form of the fork progressed from having two prongs like a modern carving fork to three and then four prongs by the end of the seventeenth century. As prongs were added, they became smaller and closer together, developing a slight curve. As a latecomer, the fork was not widely adopted by society until the nineteenth century.

In Eastern cultures, flat-bottomed spoons may be used to consume soups; however, where the fork has grown into use in the West, Asian flatware includes chopsticks instead. Chopsticks are believed to be around 5,000 years old. Although typically made of wood or bamboo, they are sometimes made of silver, ivory, or other precious materials. Because much of Asian cuisine consists of small pieces of food that do not need to be cut, the chopsticks are all that is necessary. Cuisine has historically had a great impact on the adoption of utensils. In many countries, such as India and the Philippines, the hand continues to be the conventional vehicle for moving food to the mouth.

See also Baroque; Colonial America; Guilds; Jensen, Georg; Tiffany & Company

FURTHER READING

Coffin, Sarah. *Feeding Desire: Design and the Tools of the Table, 1500–2005.* New York: Assouline and Cooper-Hewitt, National Design Museum, 2006.

Newman, Harold. *An Illustrated Dictionary of Silverware.* London: Thames and Hudson, 1987.

Van Trigt, Jan. *Cutlery from Gothic to Art Deco: The J. Hollander Collection.* Antwerp, Belgium: Pandora, 2003.

The Victoria and Albert Museum's Cutlery Collection, http://www.vam.ac.uk/collections/metalwork/metalware/cutlery/index.html

Forging

Forging is the process of applying localized force to achieve controlled deformation of a piece of metal. Typically, the force is exerted in the form of a hammer blow, and the metal being worked on rests on a hard, solid surface such as a steel anvil. On a larger scale this is achieved with mechanically driven hammers. The force of impact can be utilized to alter the shape and size of the metal's cross-section. Forging is generally used with thicker, linear pieces of metal such as rods or bars. It can be used to create a calligraphic effect in linear metal elements through transitions of thick to thin and round to flat.

Forging is one of the most ancient metalworking techniques. It is commonly associated with blacksmiths working in ferrous **metals** but may also be used in other malleable metals and at nearly any scale. Silversmiths commonly use forging as a means of producing **flatware.** Ferrous metals are usually forged while hot, which is called hot forging. Although nonferrous metals may be worked while hot, they do not localize heat as well as ferrous metals do, making them less easy to handle. For this reason, nonferrous metals are generally **annealed** and worked at room temperature, which is referred to as cold forging. Carefully forging a **cast** piece will help to realign its crystalline structure and make it less brittle.

A forging hammer usually has one flat face and one cross peen. A cross peen is a hammer face that is short but wide. The cross peen imparts pressure on a short, wide plane that pushes the metal away in the directions perpendicular to the longer sides of the hammer face. The flat face distributes force evenly and may be used for smoothing the metal's surface as in planishing (see **Raising**). Two common techniques in forging are drawing a taper and upsetting. When drawing a taper, a square cross-section is usually used. The cross peen is used with the width of the face perpendicular to the length of the rod. Overlapping blows are delivered down the length of the piece being worked. This is repeated on all sides with more repetitions being directed at the end to be made thinnest. In this way, a gradual lengthening and thinning of the metal is produced. In upsetting, a piece of metal is struck on its end or edge, forcing the metal into itself and resulting in a greater thickness where struck. These two methods were traditionally used in conjunction in order to make nails.

See also Connections; Stamping and Die Forming

FURTHER READING

Andrews, Jack. *New Edge of the Anvil: A Resource Book for the Blacksmith.* Drexel Hill, PA: Skip-Jack Press, 1994.
Artist-Blacksmiths Association of North America, http://www.abana.org

Gilding and Plating

Throughout history, metal workers have been covering one metal with the appearance of another. This is most often done with the intention of mimicking a more valued material but is sometimes also used to improve functionality. For example, before the advent of stainless steel, silver dessert spoons and forks were sometimes gilded in order that contact with acidic fruit would not cause them to tarnish (see **Flatware**). From the eighth through the twelfth centuries, gilded chalices of copper were used in the churches to impress the faithful. The goldsmithing **guilds** eventually succeeded in restricting this encroachment, and copper workers were prohibited from plating their wares.

GILDING AND LEAF

The term gilding refers to the act of decorating something with gold. When gilding is done to part but not all of an object, it is referred to as parcel gilt or party gilt. The earliest form of gilding was most likely the physical application of gold sheets, affixed with brads or tabs. As early as 1500 B.C. thinner foil, also called gold leaf, was attached with an adhesive to a variety of substrates.

The ancient Greeks used a process called fire gilding or mercury gilding. This involves making an amalgam of mercury and gold and applying it as a paste to an object made of a different metal. The object is then heated over a fire or in an oven until the mercury burns off and the gold is deposited on the surface. It was a very popular way to finish the cast bronze furniture mountings made in France at the end of the eighteenth century. In England the result of mercury gilding on bronze was called ormolu. Mercury gilding was the most widely used method of gilding before the advent of electroplating, despite its toxicity.

Depletion gilding or depletion plating is done with alloys that are mostly composed of gold or silver. The surface is treated with acids that leach out the base metals in the alloy, leaving a pure gold or silver skin. This practice has been used for many years in Mexican and South American gold work. It is also a common way to finish sterling silver in the craft of silversmithing.

Silver may be beaten into leaf form and applied as gold leaf is. However, silver is prone to react to sulfides and must be coated with varnish. Sometimes polished tin or aluminum leaf is used to replicate silver leaf.

FUSED PLATE

Around 1742 a man named Thomas Boulsover of Sheffield, England, discovered that a thin sheet of silver could be made to fuse with a thicker piece of copper and that they would stretch at the same rate. This led to the invention of fused plate, also known as old Sheffield plate, in which a thick sheet of copper was sandwiched between two thinner pieces of silver. With the addition of heat and pressure, the metals combine to form an alloy at the surface where they meet, which bonds them together. The sheet was used to make all manner of household items, resulting in a kind of silver for the middle class. By 1830, fused plate was also being made with nickel at the core instead of copper. Michael Bolton of Birmingham, England, manufactured extensively in fused plate. It was also used in France and Russia. Fused plate was popular from the middle of the eighteenth century to the middle of the nineteenth century, when its place was usurped by electroplating.

ELECTROPLATING

Electroplating uses the process of electrodeposition to build up a layer of a new metal on top of an existing form through electrolysis (see **Electroforming**). G. R. Elkington patented the process in 1840. It was most commonly used over nickel alloys, the result of which is sometimes called EPNS, or electroplated nickel silver.

See also Anodizing; Lamination Techniques; Nineteenth Century; Patina; Rococo and Neoclassical

FURTHER READING

Bury, Shirley. *Victorian Electroplate.* London: Hamlyn, 1971.
Edward Wenham. *Old Sheffield Plate.* London: Bell, 1955.
Goodison, Nicholas. *Ormolu: The Work of Matthew Boulton.* London: Phaidon, 1974.
Lewton-Brain, Charles. "Some Notes on Depletion Gilding," http://www.ganoksin.com/borisat/nenam/depletio.htm

Gold. ***See*** Metals

Guilds

Craft guilds, sometimes called mysteries, were professional organizations structured around a particular field of practice. They originated in Western Europe during the **Middle Ages** when itinerant craftsmen began to move into towns. Guilds began to develop in the eleventh century and were firmly established across Europe by the early thirteenth century.

The original purpose of the guild was to protect craftsmen through equal distribution of labor opportunities and evaluation of skills. The imposed standards of quality in the alloys that guild members used also served as protection for the consumers (see "Hallmarks" section in this entry). In some cases guilds were responsible for the medical care of their members as well. In the following centuries, guilds developed considerable political power and began to control most aspects of a craftsman's life. They regulated prices, working hours, and the number of workers in a shop; they collected taxes for the monarchy and were held responsible for the behavior of their members. There were guilds for goldsmiths, knife makers, armorers, pewterers, braziers, and most every aspect of metalworking. Goldsmith guild members worked in both silver and gold, despite the name. As the powers of the guilds grew, elements of metalworking became increasingly subdivided.

By the late fourteenth century the strict rules of the guild had become more harmful than beneficial. Skills were so subdivided that it required members of no fewer than three guilds to make a single knife, only one of whom was permitted to sell it. Increased specialization led to outstanding work in some cases but also to unemployment or lack of invention in others. The number of master craftsmen (see "The Apprenticeship System" section in this entry) was kept deliberately low to create a monopoly and artificially inflate prices. It was made as difficult as possible for journeymen to achieve master status. Some journeymen had to buy their own set of armor or pay exorbitant sums of money to become masters.

Eventually journeymen and apprentices simply became hired employees, laying the groundwork for industrialization. Attempts to break the guild systems began in the sixteenth century, and by the eighteenth century little but the social rituals of the guilds remained. The ends of most guilds were brought about by industrialization, urban growth, political instability, and their own greed. One of the rare exceptions is the still-active Worshipful Company of Goldsmiths in London, which enjoyed a resurgence of activity in the field during the twentieth century.

HALLMARKS

The kinds of marks placed on metal objects for official purposes have differed according to the needs of the time and place. The way that guilds verified the quality of metal wares was to require that finished pieces be tested and marked by the guild. Goods that did not pass the standards of purity were destroyed. This particularly applied to things wrought in gold and silver.

The collection of markings on an approved piece was called a hallmark because it was usually marked at a goldsmith's hall. The assay mark signified that the metal had been tested. The standard mark indicated the particular alloy used. The maker's mark denoted who the registered master was that took responsibility for the object. The object may not have been worked on by him directly or even have come from his shop. There are instances of marks being applied by retailers as well. In addition to these marks, metal goods may also have been marked with information about the locality of manufacture, date tested, whether taxes were paid, and the identity of the authorizing agent.

Historically, marking has been done with a small stamp, which leaves an impression when laid on the metal and struck with a hammer. Hallmarking began in France in the thirteenth century and continues to be used throughout Western Europe. The only Western country that has not had a hallmarking system is the United States, although most American goldsmiths voluntarily mark their wares and frequently include an indication of the alloy used.

THE APPRENTICESHIP SYSTEM

A number of the skilled trades utilized an apprenticeship system in the training of initiates. Although similar forms of training have been used from ancient Greece into the modern day, it was under the guilds of Europe that the apprenticeship system was most active. Prior to industrialization, most metalworkers went through some period of apprenticeship. This usually began when a young man was between 10 and 16 years of age. The more costly the material involved, the more important it was that the young man be honest and of a good family. Apprentices to the goldsmiths trade were selected very carefully.

The length of apprenticeship varied, but for a gold- or silversmiths the duration was traditionally six to eight years. The apprentice was legally indentured to a master of the trade. The novice would live in the master's house and be fed and clothed by him. In addition to the trade, apprentices were also taught to read and write. Sometimes apprentices even learned a new language. In return the boy was expected to be attentive and well-behaved and remain unmarried.

Apprentices were given menial tasks at first such as sweeping and buffing. As their skills grew they took on additional responsibilities. At the end of the apprenticeship it was expected that the individual would know enough to practice the trade independently. At this stage the trainee was called a journeyman. It was anticipated that he would spend one to three years traveling to work for different masters and continue to learn and refine his skills. After this, if his circumstances permitted, the journeyman could apply to the guild for the title of master. If he were granted permission to try for that title, he would be given the task of creating a master piece and possibly subjected to other forms of scrutiny. The master piece was a single object executed solely by the applicant under controlled circumstances and was expected to show his abilities. Once he had achieved master status, a metalworker could open his own shop and begin to train his own apprentices.

See also Armor and Arms; Arts and Crafts Movement; Byzantine Empire; Colonial America; de Lamerie, Paul; Renaissance; Stamping and Die Forming; Verdun, Nicholas of; Yellin, Samuel

FURTHER READING

The British Hallmarking Council, http://www.britishhallmarkingcouncil.gov.uk

Epstein, Steven. *Wage Labor and Guilds in Medieval Europe.* Chapel Hill: University of North Carolina Press, 1991.

Faith, Dennis. *Three Centuries of French Domestic Silver: Its Makers and Its Marks.* New York: Metropolitan Museum of Art, 1960.

The Goldsmiths Company, London, England, http://www.thegoldsmiths.co.uk

Jackson, Sir Charles James. *English Goldsmiths and Their Marks; a History of the Goldsmiths and Plate workers of England, Scotland, and Ireland,* second edition, revised and enlarged. New York: Dover Publications, 1994.

Mackenney, Richard. *Tradesmen and Traders: The World of the Guilds in Venice and Europe, c. 1250–c. 1650.* London: Croom Helm, 1987.

Nelson, Lynn H. "The Rise of Capitalism ca. 1300–1500: The Guild System." In *Lectures for a Medieval Survey.* The ORB: The On-line Reference Book for Medieval Studies, http://www.the-orb.net/textbooks/nelson/capitalism.html

Hallmarking. ***See*** Guilds

India

The country currently known as India encompasses many diverse cultures and peoples. India occupies the Indian subcontinent along with Pakistan and Bangladesh. The many different groups that have lived there have influenced the metalsmithing of the area. Although an overview will be given, it should be understood that significant regional and cultural variations exist.

The first metalworking people of the region were active from the third millennium B.C. They fashioned small copper articles and used lost wax **casting** to form bronze weapons, tools, and figures. By the first millennium B.C. the people of this area were casting iron as well as working with silver, gold, tin, and lead.

Over time, a social system that divided people into castes evolved. Castes are fixed from birth and determine one's occupation and social status. There are thousands of different castes in India, belonging to one of five levels. Metalworking castes are part of the Sudra group, the second lowest social level. Caste members may only marry and eat with others of the same level. The different castes of metalworkers are similar to the Western medieval **guild** system. Each aspect of metalsmithing concerns a different caste, such as blacksmithing or bronze deity casting. These professions are generally handed down within families. Women may aid in the work of their husbands by operating a forge, setting inlay, or polishing finished pieces. Crafts are usually practiced in or around the home.

A WHOLE LOTTA LOTA

A lota is a small vessel that is generally spherical, with a pronounced neck and flaring lip, usually made of copper or brass, which is used for carrying and pouring water in India. Some special lotas are made in half brass, half copper, to symbolize the meeting of the Ganges and Jamuna Rivers. In addition to its domestic functions, the lota is also used in religious ceremonies and Hatha Yoga. Perhaps the most interesting use of the lota form is in magic tricks. The lota is a prop in a magic act sometimes referred to as the "Waters of India." Specially made vessels seem to pour until empty, then after a few moments appear to be magically replenished! The trick was most likely pioneered by magician Kalanag (Helmut Schreiber) in the early twentieth century.

COPPER AND BRASS

In the state of Rajasthan in northwest India are deposits of copper and tin that have supplied the metalworkers of India for thousands of years. Although brass (copper and zinc) has been used since ancient times, it was made by smelting ores that combined both constituent metals. Zinc is a difficult metal to smelt alone because of its low melting temperature (see **Mining and Smelting**). India was the first to produce zinc though distillation, two centuries before the West.

India is unusual in that metal tablewares were not replaced by ceramic and glass articles until fairly recently. In India, certain metals are believed to benefit one's health when used, including brass and copper. Copper and copper alloys were used extensively for the preparation and serving of food. This has declined due to industrialization, which has brought cheaper mass produced goods, frequently of stainless steel or aluminum. Of all the traditional crafts in India, metalworking has probably declined the most as a result of industrialization.

Copper and brass vessels were usually produced with a combination of sinking and **raising.** Large hollowware pieces were pieced together in several sections, which were riveted and then reinforced with a low temperature solder (see **Connections**). Other brass articles were cast in sand and then **chased.** Some vessel forms have hardly changed in thousands of years.

Metal objects that were found throughout the houses of India would have included: oil lamps, cooking vessels, individual beakers for hand washing and drinking, bowls, cigarette boxes, vases, flasks, and locks. More expensive examples might have been decorated with **stamping, engraving, inlay, enameling,** piercing, chasing, gems encrustations, wrapped wires, or **plating.** A more recent, popular embellishment involves engraving a linear motif and then filling the recesses with brightly colored lacquers.

One metal product unique to India is the *paandaan* (also spelled *pandan* and *paan daan*). This is a box containing the tools and ingredients to make *paan,* a mixture of nuts, spices, fruit and tobacco wrapped in a betel leaf. It is chewed in the same manner, and for many of the same reasons, as chewing gum. The making of a good *paan* is still considered a valuable skill.

BRONZE

Bronze casting is done with the lost wax process. It is typically used to create religious imagery including statues of the many Hindu gods. The process of making the divine sculptures is strictly prescribed, including the proportions of the image. Processes codified in the sixth century are still used as guidelines. According to tradition, temple figures must be cast from one of two sacred alloys, *ashtadhatu* or *panchaloha. Ashtadhatu* is made up of gold, silver, copper, zinc, lead, tin, iron, and mercury. *Panchaloha* is an alloy that contains gold, silver, brass, copper, and lead and represents the five elements of earth, water, fire, air, and ether. Once the figure is cast, it is cleaned with chisels. The eyes are left for last, to allow the deity to slumber as long as possible. These works find their way into temples, wealthy houses, and distant art galleries. The cast bronze Hindu deity figures in the temple complexes of southern India are thought to shine with a divine radiance.

BIDRI

Bidri is a zinc-based alloy also containing varying amounts of tin, lead, and sometimes copper. The alloy was developed for the ease with which it receives inlay and the **patina** it takes. The word *bidri* comes form the city of Bidar, one of the main production centers of *bidri* wares. *Bidri* wares date back to at least the seventeenth century.

The production of *bidri* ware begins with the casting of the object. Once the shape has been cleaned, it is darkened with cupric sulfate. The dark background acts as a kind of layout die, providing contrast with the clean metal that is revealed through engraving. The engraving is accomplished with a hammer and varying sizes of chisels.

Once the engraving is completed, pieces of contrasting metal wires or sheets are inlayed. Traditionally the metals for inlay have been brass and silver, but gold has been used as well. When the inlay is in place and the surface has been finished, the whole piece is patinated with a chemical mixture that turns the *bidri* background metal a rich black but does not affect the inlaid metals. The *bidri* ware is then oiled, which completes the process.

The making of *bidri* wares had begun to decline in the eighteenth century, but India's participation in the international exhibitions of the nineteenth century reinvigorated the industry (see **Arts and Crafts Movement**). Objects made from *bidri* include hookahs (a kind of water pipe), bowls, vases, jugs, dishes, plates, ashtrays, cigarette cases, spitting pots, tobacco implements, and buttons.

IRON

A cast iron pillar outside of Delhi in the historic Qutb Complex exemplifies the early mastery of this material in India. It is 23 feet high and dates from the fifth century A.D. The quality of the iron is so good that it has somehow not degraded.

Contemporary blacksmiths are found either settled in urban areas or itinerant in rural areas. Itinerant blacksmiths are more common in the north as an adjunct to farming communities where they make and mend tools. Domestic utensils that have traditionally been made by a blacksmith include cooking pots and pans, scissors, knives, and ladles.

See also Ancient Rome; Art Deco; China; Filigree; First Metalsmiths; Islamic Metalwork; Lamination Techniques

FURTHER READING

Agrawala, Vasudeva Sharana. *The Heritage of Indian Art.* Delhi, India: Publications Division, Ministry of Information and Broadcasting, 1964.

Barnard, Nicholas. *Arts and Crafts of India.* London: Conran Octopus, 1993.

Cooper, Ilay, John Gillow, and Barry Dawson. *Arts and Crafts of India.* New York: Thames and Hudson, 1996.

Culture in India, http://www.culturalindia.net/indian-crafts/metal-work.html

The Freer Gallery of Art and Arthur M. Sackler Gallery, Smithsonian Institution, Washington, DC, http://www.asia.si.edu

Munsterberg, Hugo. *Art of India and Southeast Asia. Panorama of World Art.* New York: H. N. Abrams, 1970.

National Museum, New Delhi, India, http://www.nationalmuseumindia.gov.in

Stronge, Susan. *Bidri Ware: Inlaid Metalwork from India.* London: Victoria & Albert Museum, 1985.

Yule, Paul. *Metalwork of the Bronze Age in India.* Prähistorische Bronzefunde, Bd. 8. München: Beck, 1985.

Zebrowski, Mark. *Gold, Silver and Bronze from Mughal India.* London: Alexandria Press in Association with Laurence King, 1997.

Inlay

Inlay is a process by which additional materials are incorporated into the surface of a metal. The process of inlay has traditionally involved physically removing part of the surface of a hard parent metal, such as bronze or steel, with chisels or **engraving** tools. This is done in such a way as to leave an undercut in the remaining metal. The removed metal is then replaced with a visually contrasting, softer metal such as gold or silver. The new metal is hammered into place so that it expands into the undercut areas of he parent metal and is thus mechanically held in place without soldering or welding (see **Connections**). The result is sometimes referred to as *damascening,* which is not to be confused with *damascene steel* (see **Lamination Techniques**).

The traditional inlay process is most commonly used for linear motifs created with wire inlay. For larger expanses of inlay, a technique known as overlay or counter-damascening may be used. In this technique, the parent metal's surface is roughened with small, incised lines that run at right angles to one another. Over this prepared surface, thin pieces of a contrasting metal wire or foil is rubbed into the tooth of the parent metal until it is held in place by the roughened areas. With either kind of inlay, if a significant portion of the inlaid metal is allowed to protrude above the plane of the parent metal the effect may be referred to as encrusting.

There are other processes of incorporating contrasting metals into a parent metal's surface that are sometimes referred to as inlay but use soldering instead of mechanical joinery. In lamination inlay a piece of contrasting metal is soldered to the top of a parent metal and then the two are hammered or mechanically compressed until they occupy a continuous plane. In solder inlay a depression is formed in a parent metal and then solder with a lower melting temperature than the parent metal is heated until it flows and fills the recesses. The excess is removed and the result looks something like inlaid silver. In puzzle inlay or marriage of metal, an area of the parent metal is removed and a corresponding piece in a different metal is fitted and soldered into place. This differs significantly from traditional inlay in that the color change between the two metals is evident on both sides of the parent metal, rather than on just the outer surface. Metal may also be inlaid with nonmetals, such as stones or glass. In such cases, a space for the inlaid material would be cut or built within the parent metal and the new substance would be fixed in place with an adhesive.

True inlay is most commonly used in the East, especially in **Islamic, Indian,** and **Japanese** metalwork. It was found in Western Europe just prior to and during the Romanesque period of the **Middle Ages.** Inlay was not really used again in the West until around 1500, when it traveled along trade routes to Venice by Islamic craftsmen. It then was used to embellish tableware and ceremonial **armor** throughout Europe.

See also Gilding and Plating; Lamination Techniques; Niello

FURTHER READING

Smith, Cyril Stanley. *A History of Metallography: The Development of Ideas on the Structure of Metals Before 1890.* Cambridge, MA: MIT Press, 1988.

Iron. ***See*** Metals

Islamic Metalwork

Islam is a religion stemming from revelations made to the prophet Muhammad by God, called Allah, in the early seventh-century Arabia. The followers of Islam are referred to as Muslims. Their belief system quickly spread both east and west into large areas of Europe, Africa, and Asia and remains a major global religion. The time period between the fall of the Roman Empire and the development of Western European nation-states saw the geographical peak of the Muslim world. At its height, the Islamic empire surpassed the size of the Roman

Empire. Metalworking technologies developed earlier in Islamic countries than in their neighbors to the West. Through the spread of Islam and the Crusades, the Muslim knowledge of metalworking enlightened medieval Europe. In order to give a sense of traditional Islamic metalwork, this entry will deal primarily with works made prior to the seventeenth century.

The term Islamic metalwork refers to a broad category of objects that were of or for a people in one of the many places in which Islam was practiced. This means that it is not limited to a single time, place, or ethnic group. For something to be considered Islamic, it does not necessarily have to have a religious function and may not have been made by Muslims.

Islamic metalworkers exerted great skill in the manufacture of utilitarian objects. Because sculpture and painting are not major parts of the Muslim culture, much of their artistic enterprise was exerted on domestic objects. Mosques, buildings in which Muslims practice religious worship, were not patrons of the arts in the same way that Western churches have historically been.

Most traditional metalwork is not signed. Metal craftsmen trained family members, passing the trade from generation to generation. Metalworkers might have a shop in a city center or work as traveling craftsmen. In urban areas, enclaves of a certain craft would develop so that all the practitioners of a particular trade could be found in a concentrated area. This also allowed for a great deal of specialization within the trade.

METALS AND TECHNIQUES

Islamic metalwork was made with gold, silver, copper, copper alloys, iron, lead, and zinc. Most of the earlier surviving Islamic metalwork is made of copper-based alloys, such as bronze and brass. The intricately worked copper-based alloys were too attractive to destroy and not inherently precious enough to melt down, which is one reason why more of them survive.

The relative scarcity of gold and silver objects can be attributed to several factors. Historically, precious metals have not survived in their original form for the simple reason that they could be easily transformed into currency or other objects that better suited the changing tastes of the day. Also, the religion of Islam does not prescribe the use of burial objects, so no pieces were preserved as a result of having been buried with the dead. Furthermore, the Koran forbids personal ownership of precious metals. This was not always followed very strictly, and objects that were plated or inlaid with precious metals were often made as a kind of moral compromise. A form of bronze with a high tin content that came close to the appearance of silver was also used in place of precious metals.

Islamic metalwork was formed through the processes of **casting, raising,** sinking and spinning (see **Lathe Work**). Casting techniques used included open-faced as well as lost wax and sand casting. There are many fine examples of cast bronzes made prior to the eleventh century. After that period, the decorative emphasis shifted to the use of **inlay** in objects formed from sheet brass by raising or spinning. Elaborate surface decorations were attained through **chasing and repoussé,** piercing, **engraving,** setting gemstones, and **stamping.** To further complicate the designs, Islamic craftsmen used different colored wire and sheet metal inlays or created contrast in recesses with **niello** or bitumen.

IMAGERY

The revelations that Allah made to Muhammad are known as the Koran or Qur'an, and their written form is the main focus of veneration for the religion. This has led to an emphasis on the written word and book arts within Islamic artwork. Text in the form of elaborately designed scripts has been incorporated into all artistic media, including metal. In the early incorporation of text, an angular form of script called *kufic* was used. By the twelfth century, the script would become more elaborate, cursive, and intertwined with the other decorative motifs.

Islam condemns realistic representation of living things, believing that the act of creation is reserved solely for Allah. Thus Islamic craftsmen have turned to the use of abstracted figures and an emphasis on geometric motifs. Islamic metalsmiths became masters of pattern development. Common decorative motifs included geometric shapes, abstracted flora and fauna, arabesque scrolls, text, and interlacing. The arabesque is a scrolling, foliated, vinelike ornament that was commonly used in Islamic art. The fully developed arabesque was in use by the tenth century. Networks of geometric shapes were used to frame other motifs. Astrological symbolism can be found increasingly on vessels after the twelfth century.

Vegetal motifs were a principle decorative element in much Islamic metalwork. These may have been used as a kind of all-over pattern or as a single outstanding element. Specific plants that were depicted included the pomegranate and acanthus, which had also been used by the Greeks and Romans. Many Islamic countries were located along major east-west trade routes, and after the thirteenth century Chinese imagery like the lotus and peony were also incorporated.

OBJECT TYPES

The types of metal objects that were produced in Islamic regions include both domestic and religious objects. Metal was used to create architectural hardware, scientific instrumentation, armor, jewelry, vessels, utensils, candlesticks, and currency. Domestic vessels in copper-based alloys make up the bulk of surviving examples. Objects produced for the mosques were not significantly different in form than those used for everyday life, but had more elaborate decorations.

Many Muslim countries did not use eating utensils, and thus food consumption was done with the hands. Vessels for pouring water for the washing of the hands were widely made. Also, most Islamic countries are arid, making the storage and carrying of water quite important. Ewer and basin forms were prevalent and underwent many stylistic shifts.

THURIFICATION

Thurification is the scenting of a person or place through the burning of incense. This was deeply ingrained within Islamic social life and existed prior to the spread of Islam. The pre-Islamic Semitic people believed that trees were animate and sacred, and they used the burning of certain trees for ceremonial purification. Fragrant woods and aloes were burned to perfume individuals and places for religious and social reasons. At the height of the spread of Islam, elaborately decorated incense containers of brass and bronze were objects of common manufacture.

Early Islamic incense burners strongly resembled their Coptic predecessors. These had short, cylindrical bodies with hemispherical hinged lids supported by three feet. Pierced decorations allowed for the escape of smoke. This style was popular in the western parts of the Islamic world from the eleventh to fourteenth centuries.

A Persian-style hooded incense burner was popular in what is now eastern Iran during the same time period. Instead of a full dome, a cylindrical body was surmounted by a partially capped half-dome that allowed for the escape of smoke. The development of different styles of incense burners is indicative of the many different cultures that were part of the Islamic world.

DEVELOPMENT

The spread of Islam occurred so rapidly that in the early periods whatever regional style of metalwork preceded the religious conversion remained largely unchanged for quite some time. An example of this is the wonderful work in silver produced during the Sassanian dynasty in **Persia.** Vessels made for the ruling class contained elaborate scenes of the king engaged in diversions such as feasts and the hunt. These staples of Sassanian metalwork were raised, chased, repoussěd, and engraved for courtly use. Such pieces continued to be made for centuries after Persia fell to Islamic rule.

By the ninth century, some commonality of aesthetics had developed within the Muslim world. This period was marked by an increase in the use of decorative lettering. Items made for the marketplace bade good fortune to the anonymous purchaser, and commissioned works mentioned the purchaser by name and title.

The period from the twelfth to fourteenth centuries was a time of exceptional production in Islamic artwork. The variety of decorative motifs used was great. Calligraphy integrated with geometric, vegetal, and arabesque designs reached a high level of sophistication. The motifs of hunt animals that had previously been used as decoration become more imaginative and were replaced by unicorns, griffins, and similar beasts. Inlay reached a level of incredible technical achievement and complexity. This transformed lowly brass objects into works of art. Inlay was initially done with copper and then later with silver. It went from highlighting certain aspects of a design to becoming the primary decorative technique. The city of Mosul was particularly known for its inlaid wares. Because of its location on the Tigris River it was a center of trade. Mosul metalwork earned a widespread reputation, and many pieces were shipped to European rulers.

The period of exceptional inlay ended with changing metal availability and Mongol invasions of various inlay-producing regions in the thirteenth century. In its place, tinned vessels of copper-based alloys came into fashion. By the fifteenth century, decorative surface elements created with chasing, repoussé, and engraving were becoming smaller and more ornate. Designs of intricate miniaturization packed the surface of metal wares. The use of figural motifs declined.

In the sixteenth century, portions of the Islamic empire were subsumed into growing empires such as the Ottoman and Safavid. The trade that they engaged in brought new materials and increased variety to works produced in Muslim lands. In Cairo and Damascus, the use of inlay was revived to incorporate other metals with steel. This is sometimes referred to as damascening.

See also Ancient Greece; Byzantine Empire; India; Lamination Techniques; Middle Ages

FURTHER READING

Atil, Esin, and W. T. Chase. *Islamic Metalwork in the Freer Gallery of Art.* Washington, DC: The Gallery, Smithsonian Institution, 1985.

Baer, Eva. *Metalwork in Medieval Islamic Art*. Albany: State University of New York Press, 1983.

Islamic Art at the Los Angeles County Museum of Art, http://www.lacma.org/islamic_art/islamic.htm

Melikian-Chirvani, Assadullah Soure. *Islamic Metalwork from the Iranian World.* London: H. M. S. O., 1982.

Museum of the History of Science, http://www.mhs.ox.ac.uk

Smithsonian Freer Gallery of Art: Arts of the Islamic World, http://www.asia.si.edu/exhibitions/online/islamic/default.htm

Ward, Rachel M. *Islamic Metalwork.* New York: Thames and Hudson, 1993.

Japan

The early phases of metalworking in Japan were directly influenced by mainland practices found first in (present day) **China.** Over time, however, the imported technologies gave way to a uniquely Japanese aesthetic. Moreover, iron was taken much farther in the making of refined objects for use by Japanese samurai and in the tea ceremony.

In addition to iron, Japanese metalworkers also used large quantities of copper and copper alloys. Bronze was probably the first metal used in Japan. Some later copper alloys are original to Japan, including *shakudo, shibuichi,* and *sentoku* (see **Metals**). Base metals were used more often than silver or gold. The precious metals were generally reserved for **gilding,** overlays, or **inlays.**

Traditional Japanese crafts were, and in some cases still are, taught through an apprenticeship system (see **Guilds**). Although the master may select any apprentice, skills tend to be passed down through generations of the same family. This has led to the convention of a family's name being linked with fine workmanship in a particular craft, sometimes for centuries.

As an island nation, Japan's influences were limited at times, sometimes on purpose. Europeans arrived in the sixteenth century, which increased Japan's scope of trade. In the seventeenth century, however, Japan's rulers decided to cut off contact with the West, and the country remained fairly isolated until the nineteenth century. When trade resumed with Europe, the arts of this enigmatic country had

---

MINGEI

The word *mingei* literally means "art of the people" and is the Japanese term for folk art. The word was coined in 1926 by historian Dr. Sōetsu Yanagi. At the time, traditional utilitarian handcrafts that had flourished for centuries were being threatened with extinction by industrialization. *Mingei* became a national movement based on the idea that there was inherent beauty in these traditional, utilitarian objects. Supporters of the movement believe that industrially produced goods are without human qualities and are not beneficial in the same manner as a thing imbued with the spirit of human touch.

To be considered *mingei,* the article has to bear the characteristics of common, preindustrial production, including anonymity of the maker, functionality, manufacture and use in large quantities, and being endemic to the place in which it was produced. In the mindless repetition of forms, it was thought that the subservient nature of these craftsmen had produced something wholesome. These moral underpinnings regarding the effect of everyday objects on the quality of life were very similar to the Arts and Crafts Movement in Great Britain and the United States.

Dr. Yanagi and two ceramicists, Hamada Shoji and Kawai Kanjiro, traveled throughout Japan, collecting objects that met their criteria for *mingei.* Dr. Yanagi opened the Japanese Folk Craft Museum in Tokyo in 1936. Among their collections are a number of cast iron domestic vessels.

---

a significant impact on the aesthetics of Western decorative arts. The influence of Japan can be seen in the Aesthetic Movement (see **Nineteenth Century**), **art deco, art nouveau,** the **Studio Movement,** and in the work of people like Christopher Dresser (see **Arts and Crafts Movement**).

In early Japanese society, metalworkers were generally held in high regard, as they appeared to control the elements of wood, fire, metal, earth, and water. Ironworkers were in a higher class than bronze workers. This may have been because of the unsurpassed achievements of the sword smiths.

BRONZE

Japan's nearest neighbors are present-day Russia, China, North Korea, and South Korea. It was from these early mainland peoples that the technology of bronze casting came, between the third century B.C. and the third century A.D. Bronze smelting and **casting** were followed shortly by iron smelting and casting in the second century A.D. (see **Mining and Smelting**). The first objects made included weapons, tools, and *dōtaku,* which are believed to be bells. *Dōtaku* ranged in height from one to five feet, becoming larger over time. Although they appear to have no clapper and show no evidence of having been struck, it is believed that they were used as musical instruments like bells. Primarily conical with a surmounting flange and elliptical cross-section, these forms are particular to Japan. Surface designs were cast in low relief and included linear geometric motifs as well as genre scenes. These were made with the lost wax casting process. A few examples of *dōtaku* have also been found in iron.

Other cast bronze articles included mirrors, incense burners, flower vases, brush stands, and hibachi. Hibachis were vessels with vented tops that contained burning charcoal that were used for warmth or lighting pipes.

In the sixth century A.D., Buddhism arrived in Japan. Along with it came metal craftsmen from China who made cast bronze statuary, ritual implements, and shrine decorations. This included, but was not limited to, wall hangings, bells, incense burners, scripture containers, staffs, ritual vessels, and gongs. Occasionally these articles were gilded or made in iron instead. One of the most impressive feats of bronze casting is the Daibutsu (Great Buddha) of the Todai-ji Temple in Nara. Cast in parts during the eighth century, it is over 48 feet high. Buddhism became the state religion in the late seventh and early eight centuries. After two centuries, more Japanese craftsmen took up the bronze work, which was reflected in the stylistic qualities.

IRON

The first iron objects in Japan were weapons and tools. Many household items were also cast and wrought from iron, including cooking utensils, lanterns, candleholders, statuettes, sake pourers, and padlocks. Because much of Japan's history has involved civil wars, the greatest skills of the ironworker were directed at the production of **armor and arms.** In the twelfth century the aristocracy collapsed and transformed into a feudal society. The following two centuries saw the rise of the samurai, the warrior class.

Japanese armor was usually composed of a helmet, torso guard with additional attached protective pieces, shin guards, armlets, and gloves. Most of the protection was made up of small plates tied together with silk and leather cords. Hanging off of the torso would be shoulder and thigh guards. Helmets were usually made of iron plates that were riveted together, becoming smaller toward the crest. An iron mask bearing a ferocious countenance sometimes accompanied these flexible helmets. Because leather, silk, and sometimes lacquer were used in the armor, it took several craftsmen to create a full suit.

By the fifteenth century, the three-foot-long sword, or katana, was the focus of the highest levels of artistry and skill in Japan. High carbon steel pieces are stacked together and forge welded (see **Connections**). This mass is repeatedly heated, **forged** out, and folded back on itself. This kneading of the metal works out impurities and creates a more consistent carbon content throughout. The folding also creates a pattern through layering that is sometimes called damascene steel (see **Lamination Techniques**).

After this preparation, a block of steel is forge welded that contains a middle layer of a softer, low carbon steel, and exterior layers of the high carbon steel. This sandwich is forged out into the long and slightly curved blade shape. The core of softer metal provides flexibility while the harder outer layer holds a strong cutting edge. Final shaping of the blade is done by filing and grinding.

The hardening and tempering of the steel is vital to its effectiveness. A mixture of clay and other materials (a secret recipe handed down through generations) is applied to the blade with a thinner application on the blade side. The blade is heated, and when the correct temperature is reached, it is quenched in water. The timing of this is very precise. The blade, which had less of the clay mixture protecting it, will be harder than the back portion, which had cooled somewhat more slowly because of its heavier clay application.

Other craftsmen polish the blade and create its other constituent parts. The handle is made of wood, typically covered by the skin of a ray fish. This is then wrapped in silk cords. Under the silk cords may be found *menuki,* two small metal ornaments. The hand guard, or *tsuba,* is usually a round, flat disc, but it is also found with four lobes, as an octagon, square, or skull silhouette. In addition to a central hole for the sword blade, it frequently has smaller holes on either side to accommodate the handles of a knife and a needle. These smaller implements fit into the scabbard of the katana.

During the relative amity of the seventeenth through nineteenth centuries, swords were worn as indicators of status. Blades were engraved and inlaid with precious metals. *Tsuba* became a canvas for the skills of the craftsman. Early decorations were pierced, but later *tsuba* have beautiful examples of carving, inlay, overlay, **enameling, chasing,** wire interlacing, and **patination.** Different metals were juxtaposed including iron, gold, silver, *shakudo,* and *shibuichi.* Imagery could be abstract, narrative, religious, mythological, or formal. Tsuba designs were frequently based on paintings or prints. Carrying swords was outlawed for all but soldiers in 1876, and so the armorers turned their talents to other objects.

Iron was also cast into teakettles for use in the tea ceremony. Zen Buddhist monks first practiced ritual tea drinking in Japan in the twelfth century. By the fifteenth century the formalized tea ceremony had become a common social gathering for the affluent, focusing on the aesthetics of simplicity. The highly ritualized activities of the ceremony are calculated to produce meaningful contemplation of the rustic and natural. Tea kettles for the ceremony were sand cast with designs in low relief. Motifs were frequently inspired by paintings and incorporated natural imagery.

ENAMEL

Japanese artisans used both cloisonné, champlevé, and plique-à-jour enameling techniques (see **Enamel**). Like the precious metals, enamel was used sparingly as decoration. When trade with the West was resumed in the nineteenth century, exported Japanese enamels were in great demand in other countries. By the twentieth century, Japanese cloisonné had become a significant export.

LACQUER

Lacquer wares were made in China first. However, so much lacquer ware was exported from Japan, Westerners began to refer to the style of finish as Japanning. Lacquer is made from the sap of certain trees found in south and southeast Asia. Once the sap is strained and its water content cooked off, it is colored with dyes. Lacquer must be applied to its substrate in many very thin layers and allowed to dry in warm and humid conditions. This slow process ensures that the whole layer is dry, not just the top skin. Each layer is polished with a mild abrasive before the next coat is applied. Dozens or hundred of layers are built up over wood, metal, and ceramic bases.

Lacquer ware resists water, corrosion, some acids, moderate heat, and insulates from electricity. Typical lacquer articles include tablewares and toiletries. There are many ways of decorating lacquer ware, including carving, inlay, and painting. The highest achievement of lacquer work in Japan is *maki-e,* or sprinkled picture. In *maki-e,* gold, silver, and their alloys are embedded in the lacquer. Great subtlety is achieved through the variety of the alloys and the different sized particles into which the foil has been cut.

See also Armor and Arms; Art Deco; India; Schwarcz, June

FURTHER READING

Association for the Promotion of Traditional Craft Industries, Traditional Crafts of Japan, http://www.kougei.or.jp/english

Feddersen, Martin. *Japanese Decorative Art.* Katherine Watson, trans. New York: Thomas Yoseloff, 1962.
Hutt, Julia. *Understanding Far Eastern Art: A Complete Guide to the Arts of China, Japan and Korea—Ceramics, Sculpture, Painting, Prints, Lacquer, Textiles and Metalwork.* Oxford: Phaidon, 1987.
Nihon Mingeikan (Japanese Folk Craft Museum), http://www.mingeikan.or.jp
Ogawa, Masataka. *The Enduring Crafts of Japan, 33 Living National Treasures.* New York: Walker/Weatherhill, 1968.
Robinson, B. W. *Japanese Sword-Fittings and Associated Metalwork.* Geneva: The Baur Collection, 1980.
The Society for the Preservation of the Japanese Sword (NBTHK), http://www.nbthk.net/NBTHKe/NBTHK.html
Williams, Marjorie. *In the Nature of Materials: Japanese Decorative Arts.* Cleveland, OH: The Cleveland Museum of Art, 1977.
Yanagi, Muneyoshi, and Bernard Leach. *The Unknown Craftsman, a Japanese Insight into Beauty.* Tokyo: Kodansha International, 1972.

Jensen, Georg
(1866–1935)

Georg Jensen was a Danish silversmith who had dreamed of being a sculptor. Instead, he founded a silversmithing workshop that was to be of international scale and influence. The innovative jewelry, flatware, and hollowware produced under his name brought Danish silver to the attention of the world.

In the late nineteenth and early twentieth centuries the British **Arts and Crafts Movement** as well as the burgeoning **art nouveau** style were changing the decorative arts all over Europe. In Denmark the equivalent movement was called *Skønverk,* an amalgam of the words beautiful and work. Danish silver from this period became widely popular in Europe and America, and the style was synonymous with the name of Georg Jensen.

Although the Georg Jensen style was the cooperative work of a number of designers, the direction originated from the hands of Jensen himself. He was born in the industrial town of Rådvad, just north of Copenhagen, where his father worked in a knife factory. As the seventh of eight children, Georg was also expected to work in the factory to help support his family. In these formative years, Georg showed technical facility at his employment and an aesthetic interest inspired by the woodlands near his home. Even as a young man, Georg wanted to be a sculptor, and his first figures were modeled from local clay that he dug himself.

In 1880 the family moved to Copenhagen and Georg was apprenticed to goldsmith A. R. Andersen (see **Guilds**). During these four years of training

Jensen augmented his education with classes in drawing, modeling, and **engraving.** By 1884 Georg had completed his apprenticeship and began to work as a goldsmith by day while continuing to sculpt in clay at night. In 1887 he was accepted to the sculpture program of the Royal Danish Academy of Fine Arts in Copenhagen. In 1891 he married, and in 1892 he graduated. By 1895 he had two sons to feed, and although he continued to exhibit his sculptural work, it was goldsmithing that paid the bills.

The years between 1892 and 1904 were eventful for Georg. He made models for a few porcelain manufacturers and operated an independent pottery with Christian Joachim Petersen for a time. He also received travel grants that allowed him to visit France and Italy. His first wife passed away, and he remarried and had a daughter. Whenever he needed income he was able to fall back on his goldsmith's training.

Georg had been executing and exhibiting metalwork of his own design since 1899, and in 1904 decided to open his own shop and pursue the craft on a full-time basis. Georg rented a small, second-floor shop with money borrowed from a Copenhagen businessman. Instead of a showroom, Jensen had a small wooden and glass display case that he hung outside at street-level every day. His first products were jewelry, as it required less material investment. He also chose to work in silver and semiprecious stones rather than gold or other expensive materials. After a successful first year he began to make hollowware and **flatware** as well.

The workshop's output was well received. It was affordable to the rising middle class, and the style was contemporary with an air of classicism. The Danish Museum of Decorative Art acquired several pieces for its collection, which increased Georg's popularity in Denmark. The work was so innovative at the time that it was soon appreciated abroad as well. Initially, Germany was the main importer of Jensen's work. In 1905 art patron Karl Ernst Osthaus organized an exhibition of Jensen's work in Germany at the Folkwang Museum in Hagen. Carl Dyhr opened a retail shop that carried Jensen's work in Berlin in 1909.

As time went on, the reputation of Jensen silver grew. The work won a gold medal at the Brussels International Exhibition in 1910. In 1914 one of his pieces was acquired by the Decorative Arts Museum in Paris. When sales in Germany halted because of World War I, Swedish consumers more than filled the gap. Between 1918 and 1925 retail shops were opened in Paris, London, New York, and Barcelona. In the 1920s and 1930s America was a leading market for Jensen exports. The popularity of Danish silver was reflected stylistically in the early years of the **Studio Movement.**

These successes were not Georg's alone, however. At first Jensen employed a journeyman goldsmith named Otto Strange Friis. Soon after, he took on apprentice Henry Pilstrup, who spent many decades with the company and eventually became a foreman of the workshop. Georg was a prodigious designer but also welcomed collaborations with his associates from the art world. Because of his particular background, Georg was able to bridge the two fields of fine art and technical goldsmithing, which meant that those inexperienced in metalworking could provide plans for silver wares, which Georg's workshop could execute. The characteristic simplicity of tablewares produced under the Jensen name was consistent, but each person who contributed brought his own style to it. If the size of the object permitted, the identity of its designer was marked upon it. Collaborators were always given credit for their designs at exhibitions. The many artists associated with the firm over the years included Mogen Ballin, Gundar Albertus, Harald Nielsen, Henning Koppel, and Sigvard Bernadotte. Two of Georg's sons were also designers for the company, Jorgen Jensen and Søren Georg Jensen.

The most significant and lasting collaborative relationship for Georg Jensen was with Johan Rohde. Rohde was a painter who was active in the modern Danish art movement. In 1906 Rohde approached Jensen to execute a design for holloware Rohde had devised. This became the fist of many such collaborations, and Rohde continued to provide designs for Jensen and other silver firms. He signed a formal contract with Georg Jensen in 1917 and the two were friends and colleagues for the remainder of their lives. The tradition of partnerships has continued throughout the life of the company.

Georg Jensen's period of success was not without its problems. A few years after going into business, Jensen's second wife died, and he again remarried and had three more children. Although the staff had grown to nine people by 1908, the workshop was using slow, traditional methods of manufacture that the sales prices could not justify. In 1912 the workshop moved to a larger space and opened their first dedicated retail shop in Copenhagen. World War I caused a decline in profit, and in order to raise funds for expansion the workshop was turned into a public limited company called Georg Jensens Sølvsmesie A/S. Georg held the majority of shares and remained director. Further expansions and financial troubles led to Georg stepping down as director and losing majority shareholder status. The company divided into separate concerns for local sales and manufacture, respectively. Georg remained artistic director of manufacture for the rest of his life, becoming more distanced from the company toward the end.

The company remained in business after Georg passed. It was subsumed under the Royal Porcelain Manufactory PLC in 1972, which became Royal Copenhagen PLC in 1985. Since 1997 they have been under the parent company of Royal Scandinavia A/S.

The impact of Georg Jensen silver on twentieth-century decorative arts was wide ranging. Its popularity was grounded in excellent workmanship and distinctive styling. Jensen and the other designers had broken with traditional historical forms. The new designs were clean and showed off the natural glow of silver. Common motifs included volumetric expanses without adornment surmounted by small stylized bunches of fruit or flowers as carefully orchestrated focal points. Some later designs were ornamented only by the subtle transitions of the **raised** form itself. Jensen flatware tended to have shorter knives and more widely spaced fork tines. The finish that Jensen was best known for was achieved by applying a heat patina, dipping the piece in sulfuric acid, and then lightly buffing the etched surface. This imparted the warm glow of well-cared-for silver heirlooms. Georg Jensen silver was a moderate form of **modernism** that was agreeable to a broad constituency.

See also Art Deco; Bauhaus; Nineteenth Century; Patina; Schwarcz, June; Tiffany & Company

FURTHER READING

Bröhan-Museum, Dedo von Kerssenbrock-Krosigk, and Claudia Kanowski. *Modern Art of Metalwork: Bröhan-Museum, State Museum of Art Noveau, Art Deco and Functionalism (1889–1939), Berlin.* Berlin: The Museum, 2001.

The Danish Museum of Art and Design, http://kunstindustrimuseet.dk/en

Drucker, Janet. *Georg Jensen, a Tradition of Splendid Silver.* Atglen, PA: Schiffer Publishing, 1997.

Georg Jensen (company Web site), http://www.georgjensen.com

Georg Jensen Sølvsmedie, A/S. *Fifty Years of Danish Silver in the Georg Jensen Tradition.* Copenhagen, Denmark: Dyva & Jeppesens Bogtrykkeri A/S, 1955.

Renwick Gallery of the National Collection of Fine Arts. *Georg Jensen Silversmithy: 77 Artists, 75 Years.* Washington, DC: Smithsonian Institution Press, 1980.

Judaica

The Jews are a people descended from the ancient Hebrews. Originally they lived in Israel, but successive invasions drove them across Europe and Asia. The Jews have also suffered religious persecution as a minority in many countries and are now scattered throughout the world. What they share is the religion and lifestyle that is Judaism. Judaism was the first religion that taught belief in a single God and is the oldest of the major world religions.

Judaica are the implements that are used in the practice of the customs of Judaism. Judaism encourages the making and use of beautiful objects for ceremonies. This is rooted in the Jewish concept of *hiddur mitzvah,* which refers to the beautification of the commandments and reflects the belief that performing religious acts should be joyful.

The first ceremonial implements made were for the original Jewish Temple, the Temple of Jerusalem. Those implements became the model for all those that followed. This tendency to look back on the original manifestations of these forms has given Judaica a certain amount of consistency. Simultaneously, as the Jewish people came to occupy many different countries, the decorative style of such wares reflected the fashions of the times and places in which they were made. Most surviving Jewish ceremonial implements date to the seventeenth century or later, due to the recurrent exile that the Jews faced.

In **Islamic** countries, the decorative motifs on Judaica tended to be more abstract and include the use of **inlay.** During the late **Middle Ages,** decoration reflected the Gothic art of the period. Judaica likewise picked up the **baroque** style during the seventeenth century. The baroque corresponded with a period of relative peace for the European Jewish community; hence there were a lot of pieces commissioned in that style. Because of these stylistic similarities to the other metal wares of their times, Judaica is sometimes best recognized by the presence of Hebrew inscriptions.

In the Islamic world, the prohibitions against precious metal objects may have kept many Muslims out of the trade of silversmithing, so it is likely that some Jewish practitioners existed. However, Jews were not admitted to the guilds of Western Europe, so until the nineteenth century Christians made most Western Judaica. There were a few kinds of work, however, that Jewish craftsmen could engage in because they were of less interest to the general population. These included working with pewter, making **filigree,** and using **niello.** The nineteenth century saw a greater participation in the design and construction of fine metal Judaica by Jews themselves, and the personal insight that they brought to ceremonial objects has evolved the genre even further.

Jewish ceremonies take place both in public at the synagogue, a Jewish house of worship, and in one's own home. In the synagogue the most important objects of Judaica are those connected with the Torah scrolls. The central text of Judaism is the Torah, a scroll containing the first five books of the Hebrew Bible. The Torah may be housed in a wooden cabinet called an *ark,* which in the West is traditionally placed against the eastern wall of a synagogue so that those facing it are also facing toward Jerusalem. In Islamic countries, the Torah itself is encased in a *tik,* or case, which might be made out of metal, leather, or wood. Most Western synagogues wrap the Torah in some kind of fine cloth. Added to this may be decorative finials on the ends of the Torah staves, bells, a crown, a hanging breastplate, and a pointer, all of which would have been fashioned out of the finest metals that the community could afford.

The tradition of decorative finials comes from accounts of the original Torah from ancient times. Individual finials or crowns on the two staves are called *rimon.* This is the Hebrew word for pomegranate, and many of the older *rimon* reflect the shape of that fruit. In the seventeenth century, they were shaped like medieval towers. A single large crown on both staves is known as a *keter Torah.* The crown form did not develop until the Middles Ages. Stave decorations frequently have bells included in them to tell the congregation when the Torah is being used.

Hanging over the staves like a necklace is the *tas,* a decorative breastplate or shield, which might describe whatever festival is being observed at the time. A hand-shaped Torah pointer, known as a *yad,* hangs over a single stave. The *yad* and *tas* were added over time and were not a part of the original ancient Torah display. The "dressed" Torah is a magnificent site. The elements are usually executed in silver or silver gilt.

MEZUZOT

Mezuzah is Hebrew for "doorpost," and its plural form is *mezuzot.* A mezuzah is a scroll inscribed with portions of the bible known as *Shema Israel* or Hear Israel. This is made up of passages from Deuteronomy 6:4–9 and Deuteronomy 11:13–21. The name of God is written on the back. Jewish households have this scroll affixed to the right side of their doorways. Displaying a declaration of one's faith on the entrance to a house has ancient roots in many cultures, including Egyptian and Islamic.

The paper scroll was originally affixed directly to the doorway, and eventually niches in doorways were created to hold it. Some doorway niches had ornamental silver coverings to protect the scroll. Those evolved into special cases made for attachment to the doorway. These come in a variety of shapes, sizes, and materials but are generally the same dimensions as a cigarette lighter. Exquisite versions in metalwork have developed since the eighteenth century that are collected by Judaica enthusiasts.

The Jewish home will also have implements for weekly and seasonal observances. These may include candlesticks, candelabrum, an oil lamp, special plates, a saltcellar, an alms box, cups, and a spice box. The Sabbath, which takes place from sundown on Friday to sundown on Saturday every week, is a prescribed day of rest with restrictions on activities. The ceremony that starts the Sabbath utilizes a candlestick or lamp and a kiddush cup. A family may also have a special Sabbath plate and an alms box. The Sabbath lamp evolved from an oil-burning ceramic kind that was used every day in ancient times to far more elaborate, many-wicked metal versions. In some countries it has been replaced entirely by candles. Sabbath ends with an event called *havdalah,* which similarly uses a light source and wine cup, with the addition of a spice box. Special vessels known as *kiddush* cups are used when the taking of wine is part of a Jewish religious observance. A few have been made in glass, but most are made of silver.

Although the plates and cups for rituals may resemble ordinary household wares in form, they are set apart by decorative motifs. These may be made of metal, ceramics, or glass. In metal, they are found in silver, pewter, and brass. Decoration is usually achieved through **chasing and repoussé, etching,** or **engraving.**

The spice box, or *bsamim,* used in *havdalah* has taken many sculptural forms. Because spices used to be dried and kept in towers, tower-shaped boxes can be found in Western Europe. Polish spice boxes tend to be more figurative, resembling a person or animal. Spice boxes from Islamic lands usually have floral shapes. Of course, there are also simple and discreet rectangular boxes used.

Another Jewish celebration that takes place in the home is Hanukkah, the Jewish festival of lights, which is celebrated for eight consecutive days each winter. The focal object of this celebration is the Hanukkah lamp or menorah. When the Jews recovered the Temple of Jerusalem in 165 B.C. after the Syrians had desecrated it, they sought to relight the everlasting lamp that had been extinguished. Although there was only enough oil remaining to light the lamp for one day, the lamp burned for eight days, which was long enough to acquire more oil and keep the flame burning. This miracle of the lamp is remembered by the lighting of the menorah. It consists of eight lights, one for each day of the miraculous burning lamp, and a *shamash,* or servant light, which is used to light the others. Ancient menorahs descended stylistically from Roman oil lamps. In some versions the lights are freestanding, as in a candelabrum. In other examples, such as typical German and Polish examples, there is a back plate that reflects the light. These are made of metal, whereas the kind used in Yemen is carved from a block of stone.

In addition to the previously mentioned forms, a Jewish household might also have plates and accoutrements for a variety of other festivals, including Seder. Seder is the meal that celebrates Passover and commemorates the Jewish flight from Egypt. The Seder plate is central to the meal and might be made of any number of materials.

See also Persia

FURTHER READING

Berger, Maurice, and Joan Rosenbaum. *Masterworks of the Jewish Museum.* New York : Jewish Museum, 2004.

Fishof, Iris, ed. *Jewish Art Masterpieces from the Israel Museum, Jerusalem.* Southport, CT: H. L. Levin Associates, 1994.

Furman, Jacobo. *Treasures of Jewish Art: From the Jacobo and Asea Furman Collection of Judaica.* New York: Hugh Lauter Levin Associates, 1997.

Grafman, Rafi. *Crowning Glory: Silver Torah Ornaments of the Jewish Museum, New York.* New York: Jewish Museum, New York, under the auspices of the Jewish Theological Seminary of American, 1996.

Jewish Museum, Berlin, Germany, http://www.juedisches-museum-berlin.de/site/EN/homepage.php

Jewish Museum, London, UK, http://www.jewishmuseum.org.uk

The Jewish Museum, New York, http://www.jewishmuseum.org

Kanof, Abram. *Jewish Ceremonial Art and Religious Observance.* New York: Abrams, 1970.

Sed-Rajna, Gabrielle. *Jewish Art.* Sara Friedman and Mira Reich, trans. New York: H. N. Abrams, 1997.

Lamination Techniques

There are a few techniques that utilize the ability of metals to form bonds with one another through surface alloying. This makes it possible to join two different metals without solder (see **Connections**) while preserving their different appearances.

MOKUMÉ GANE

Mokumé gane is Japanese for "wood grain metal." This describes the most common pattern generated with this lamination technique. In *mokumé gane,* layers of contrasting metals are stacked together, usually to a depth of 8 to 20 layers. The stacked layers are known as a billet. Under pressure, the billet is heated in a reducing atmosphere. A reducing atmosphere is a fuel-rich combustion that consumes any oxygen present, preventing the formation of metallic oxides. The billet must reach a temperature at which the different metals exchange molecules between their surfaces without melting entirely. After this exchange (while still hot) the billet is **forged** to reduce the thickness and improve the molecular grain structure of the metals.

The finished billet forms the canvas for patterning. There are a number of techniques for patterning *mokumé gane.* The most commonly used patterning techniques involve cutting into the billet to expose the many layers, then forging the billet down to an even thickness. This produces a flat sheet of metal that exhibits the contrasting colors of the billet metals, usually in an organic design reminiscent of wood grain. Heating for *mokumé gane* was traditionally done in a charcoal forge, but modern heating apparatuses such as gas-fired kilns have also been adopted for *mokumé gane* production.

The art of *mokumé gane* comes from Japan, where Denbei Shoami invented it. It has been practiced there since the late seventeenth century and was developed for the making of sword elements. The making of *mokumé gane* declined over the following two centuries and was nearly lost during industrialization. During the twentieth century, interest in *mokumé gane* revived in both Japan and the West during the **Studio Movement.** The fusion that takes place during the making of *mokumé gane* is the same as that which made Sheffield Plate possible (see **Gilding and Plating**).

DAMASCENE STEEL

The term damascene steel refers to a ferrous material (see **Metals**) that has a visible constituency of two or more elements. This material is also sometimes referred to as Damascus steel or watered steel. The name comes from the city of Damascus, in present-day Syria, which was historically a major trade center from which Western Europe acquired such metals. There are two methods of creating the appearance of this variegated material, by **casting** or by forge welding. Both techniques are believed to have originated in **India** and the Middle East. Damascene steel is different from the **inlay** process of damascening.

Cast damascene steel is sometimes called wootz. This is made by melting pieces of iron, steel, and charcoal in a reducing atmosphere. Once the carbon from the charcoal had been sufficiently absorbed by the metals, the mixture is cooled very slowly. This produces a visible crystallized structure of variable carbide content. Forging this product into a desired shape alters the observable pattern and creates the look of waves, or watering.

In fabricated damascene steel, sometimes called pattern welded steel, two or more linear elements of iron and steel are laid on one another and forge welded together. The number of layers may be increased by forging out the length of the composite, folding it back onto itself, and then repeating the welding process. The resulting laminate may be manipulated in a number of ways in order to develop different patterns with the contrasting layers.

The function of both these processes was twofold: to produce steel that was harder and more malleable than cast iron, and to create a beautiful object. These concerns were most important to the making of bladed weapons such as swords (see **Armor and Arms**).

See also First Metalsmiths; Islamic Metalwork; Japan; Mining and Smelting; Persia

FURTHER READING

Figiel, Leo S. *On Damascus Steel.* New York: Atlantis Arts Press, 1991.
Gowland, William. *Metals and Metal-working in Old Japan*. London: s.n., 1915.
Midgett, Steve. *Mokume Gane: A Comprehensive Study.* Franklin, NC: Earthshine Press, 2000.
Sachse, Manfred. *Damascus Steel, Myth, History, Technology Applications.* Düsseldorf, Germany: Stahleisen, 1994.

Lathe Work

The lathe is an ancient tool most likely descended from the bow drill. The two primary kinds of lathes used by metalworkers are turning lathes and spinning lathes. Both systems are generally oriented on a horizontal axis. In each case a power source turns a central element to which the thing being worked on is affixed.

TURNING LATHE

The turning lathe, sometimes called a machining lathe, was one of the most complex machines of antiquity. Variations on this technology have been used in the fields of woodworking, ceramics, and metalworking. There is evidence of metal turning dating back to the Bronze Age (see **First Metalsmiths**). The earliest turning lathes rotated in two directions, creating a reciprocal motion. They were powered by a cord wrapped around one of the moving elements and affixed at one end to a resisting force, such as a springy pole. The other end of the cord was attached to a foot-operated lever, which was pressed and released against the tension of the pole, causing the lathe element to spin. Continuous drive lathes, which rotated in only one direction, came into use in the fifteenth century.

Turning is essentially a reductive process. A tool with a sharpened edge is held against the spinning metal, shaving off the topmost layer by degrees. Frequently some kind of lubricant is used to facilitate the process. Different tooling may be used to incise into the end of the spinning materials to create interior voids in a process known as boring. Both of these techniques were generally used on nonferrous metals (see **Metals**) until the eighteenth century. With industrialization came improved power sources and lathes built sturdily enough to permit the turning of ferrous metals. The symmetrical precision thereby achieved contributed to improvements in large-scale production of parts for other machines.

Threaded hardware, such as screws, was historically made on turning lathes (see **Connections**). Pewterers also used this system to scrape and polish their wares after **casting** (see **Colonial America**). It is easier to turn metals that cut cleanly such as brass. Soft metals, such as copper, are harder to turn.

SPINNING LATHE

In spinning, a disc of metal is held at its center against a form called a *chuck* while they both rotate toward the metalworker. As the lathe spins, tools are used to compress the metal against the chuck. This is done from the center outward, always working below the midpoint at which the object is held. The tools used resemble long, blunt steel fingers, usually set into wooden handles, several feet in total length. The shapes of their ends vary, and many spinners customize these. The spinning tools are situated against a tool rest that can be fixed in different locations in order to provide a fulcrum point. The wooden handle is usually tucked under the right arm of the spinner.

A chuck is a solid form made of hardwood, brass, or steel. The form of the chuck represents what the interior of the finished, spun form will be. Spun forms are usually axially symmetrical, although some special lathes can spin ovoid forms. Complex hollow wares with undercuts are either spun in sections and assembled, or manufactured using a puzzle chuck or sectional chuck, which is

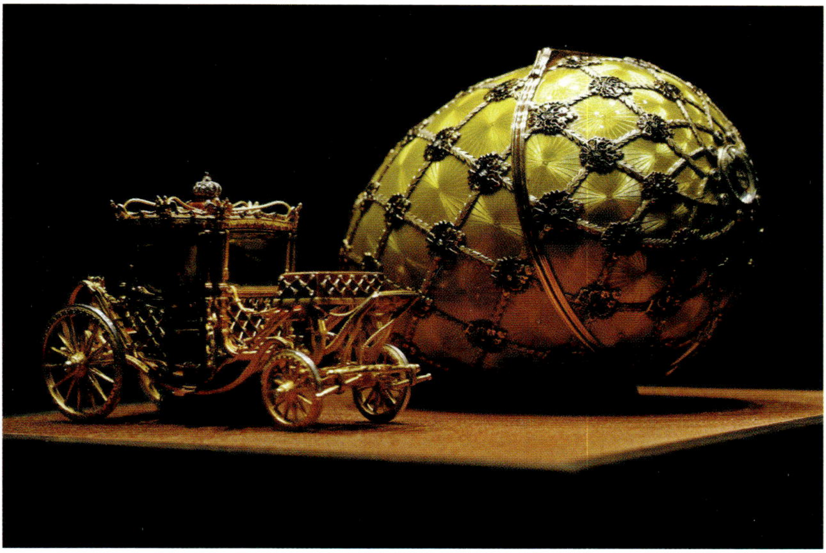

The Coronation Egg from the jewel house of Peter Carl Fabergé. Presented by Czar Nicholas II to his wife, Czarina Alexandra Feodorovna for Easter in 1897. The egg holds a perfect replica of the coach Czarina Alexandra rode into Moscow in 1897. Materials include gold, enamel, diamonds, rubies, and rock crystal. Height of egg is 5 inches (12.7 cm). AP Photo/KEYSTONE/Alessandro Della Bella.

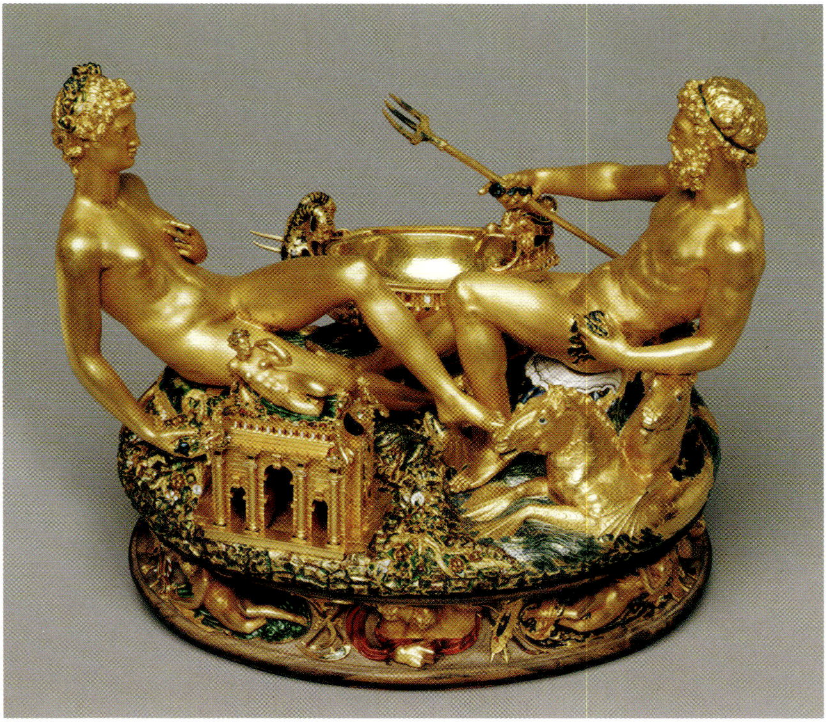

Saltcellar by Benvenuto Cellini executed for King François I of France by 1544. Personifications of the sea and earth are seated alongside receptacles for salt and pepper, respectively. Cast and hammer-formed gold and enamel on an ebony base, 10 × 13 inches (26 × 33.5 cm). AP Photo/Kunsthistorisches Museum.

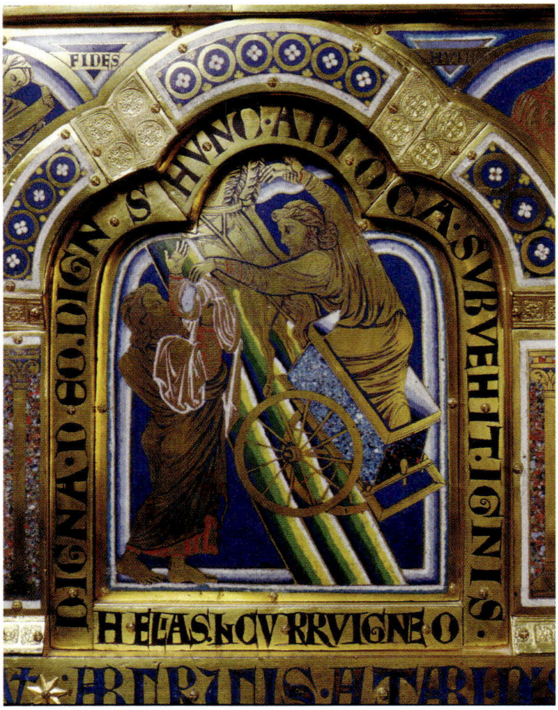

Ascension of Elijah from the Nicholas of Verdun Altarpiece, champlevé enamel on gilt copper, completed 1181, Mosan. Height is 8 inches (20 cm). The Art Archive/Klosterneuburg Monastery Austria/Alfredo Dagli Orti.

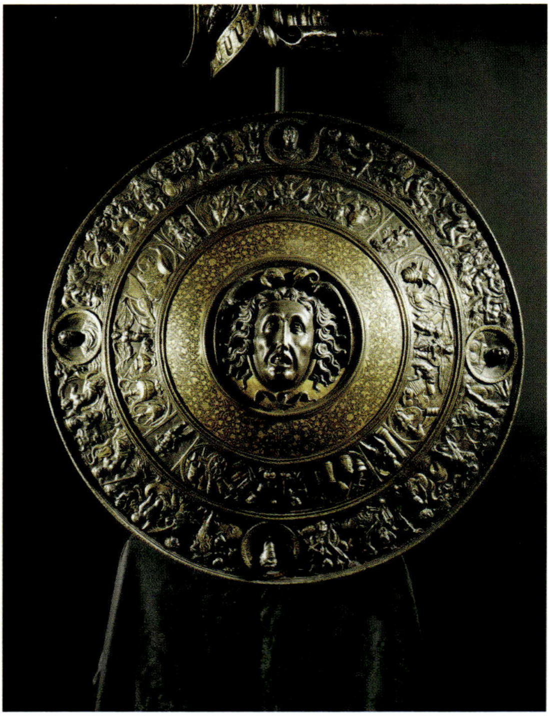

Embossed pageant shield with Medusa head from 1541. This was a gift to Emperor Karl V from his brother Ferdinand, in memory of the emperor's victorious campaign against the Turks in Tunis. Blued steel, embossed with gold ornaments, 24 inches (61 cm) across. Erich Lessing/Art Resource, NY.

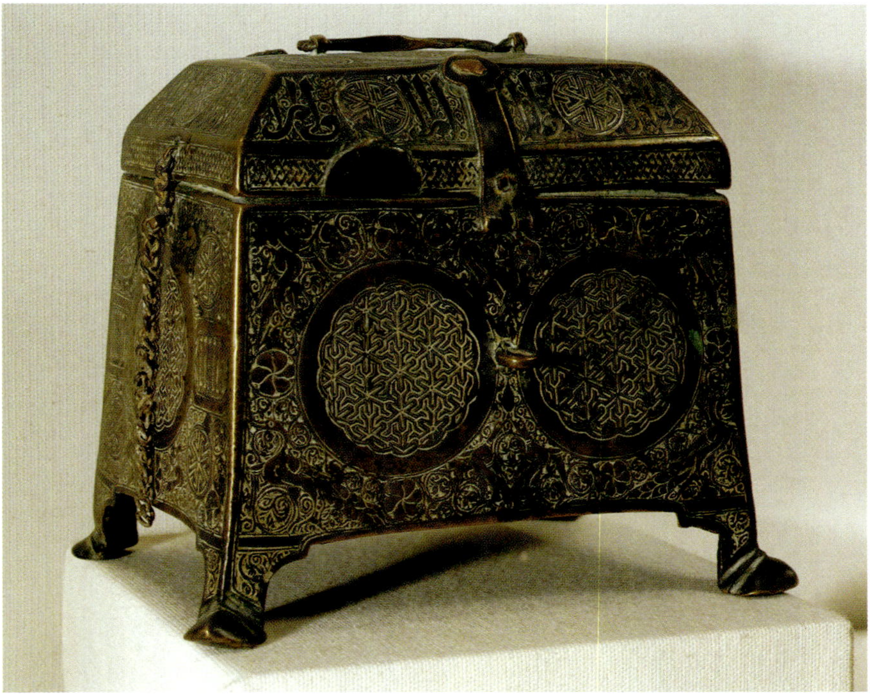

Brass box with silver inlay, Seljuk, thirteenth century. The Art Archive/Turkish and Islamic Art Museum Istanbul/Gianni Dagli Orti.

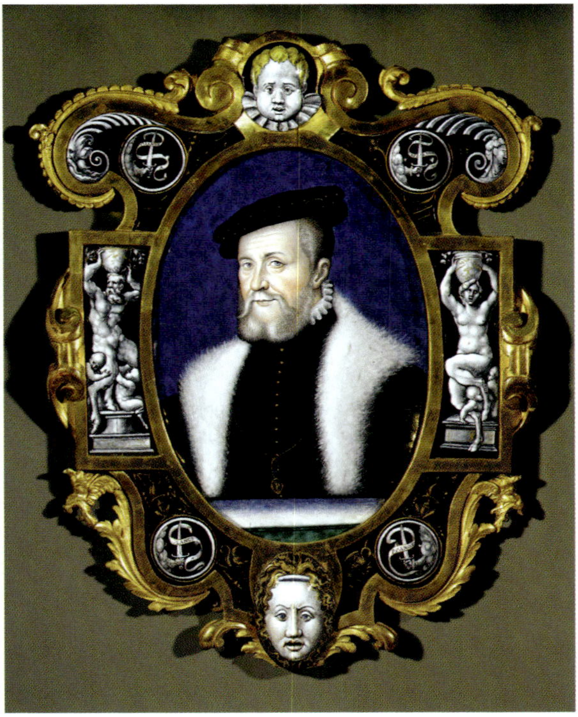

An enamel portrait of the constable of Montmorency executed by Leonard Limosin of Limoges, France in 1556. The central portrait is a fine example of painted enamels while the six surrounding plaques exhibit the *en grisaille* technique. The enamel is on copper, mounted on a gilded wood frame. Height 28.3 inches (72 cm). Louvre, Paris, France. Erich Lessing/Art Resource, NY.

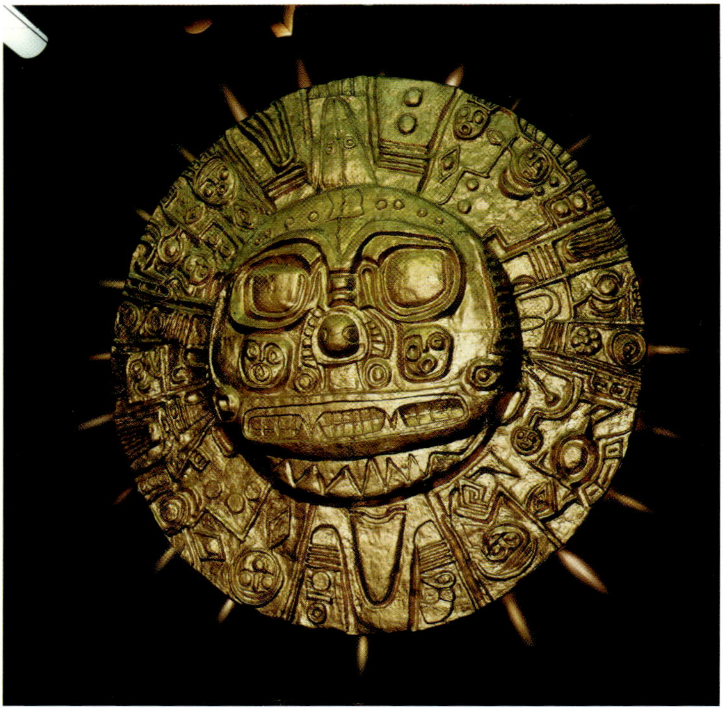

Gold Inca sun plate in National Museum (Museo de la Nacion), Lima, Peru. The Art Archive/Stephanie Colasanti.

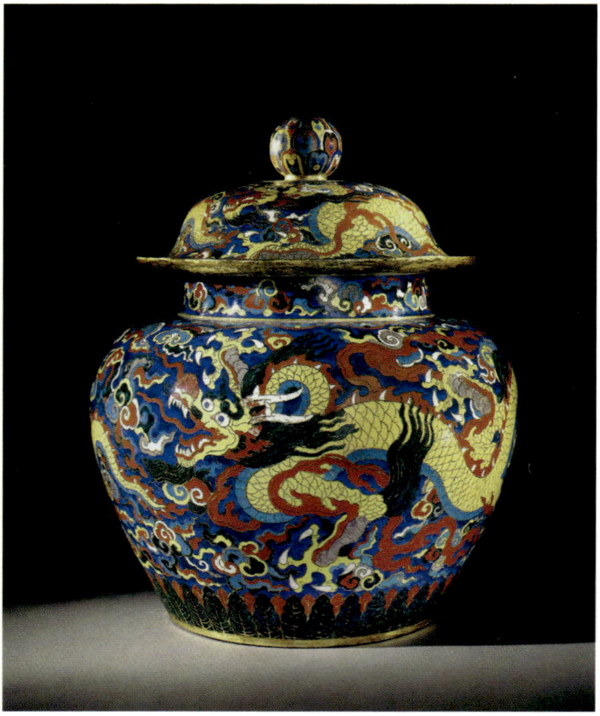

Cloisonné jar, Ming dynasty, Xuande period, ca. 1426–ca. 1435, depicting a dragon among clouds. The inscription on the neck of the jar shows that it was made under the auspices of the Yuyongjian, a division of the imperial household. The Chinese perfected the cloisonné enamel technique in the fifteenth century. By the time this jar was made it was considered appropriate for imperial use, and many superb pieces were made for palaces and temples. Height is 24.5 inches (62 cm) Erich Lessing/Art Resource, NY.

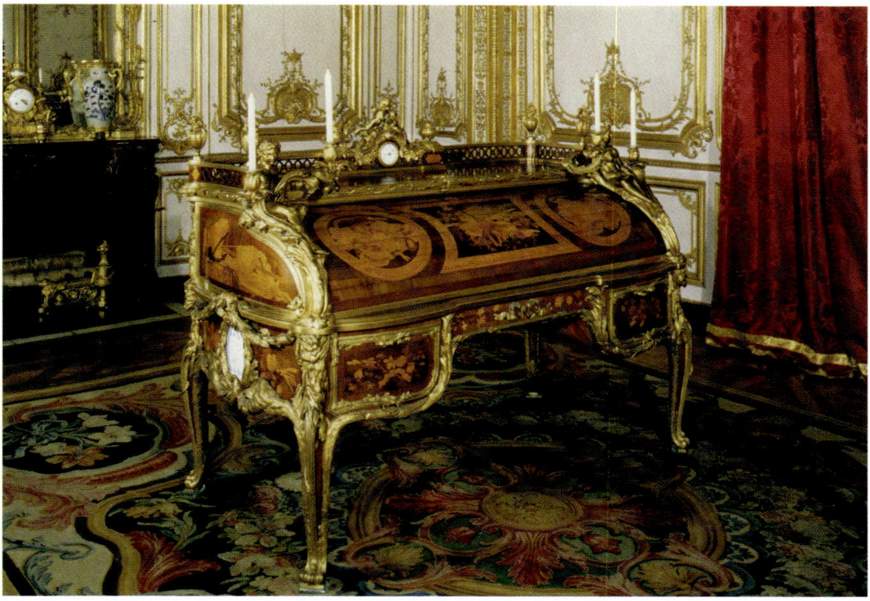

The roll-top secretary of Louis XV was begun by Jean Francois Oeben and finished by Jean-Henri Riesener in 1769. The oak frame and intricate wood veneers are framed by cast and gilded bronze mounts. Chateaux de Versailles et de Trianon, Versailles, France. Réunion des Musées Nationaux/Art Resource, NY.

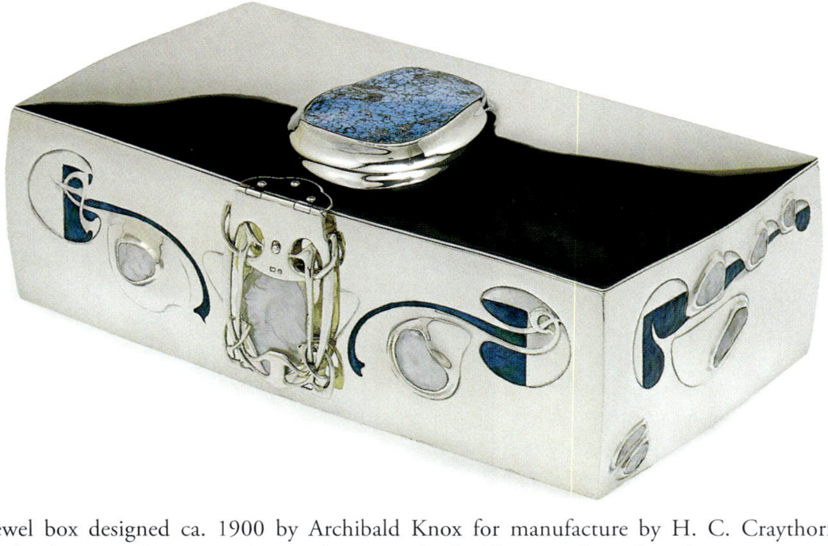

Jewel box designed ca. 1900 by Archibald Knox for manufacture by H. C. Craythorne and Liberty & Company, Birmingham, England. Silver, mother-of-pearl, turquoise, and enamel. Length: 11.5 inches (29.2 cm). Gift of the family of Blanchette Hooker Rockefeller. (327.1949). © The Museum of Modern Art/Licensed by SCALA/Art Resource, NY.

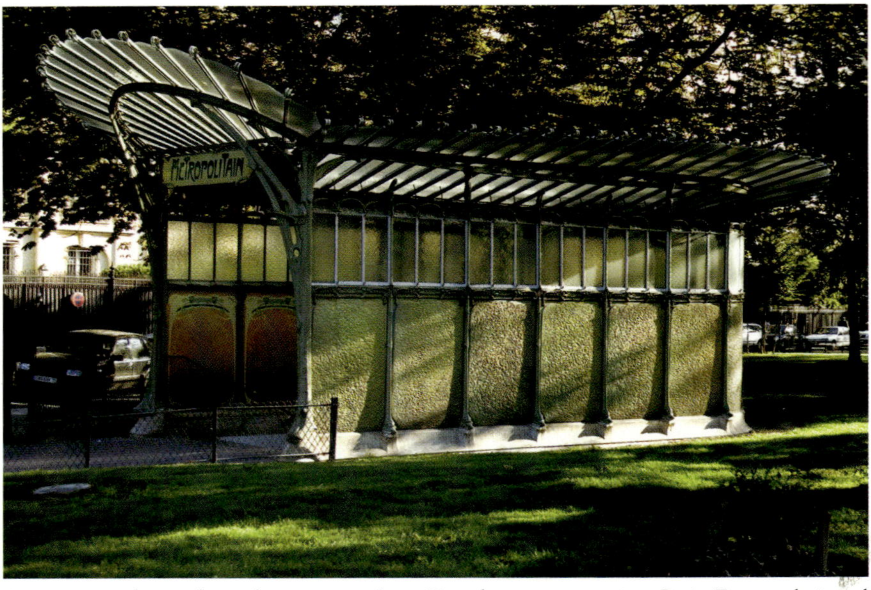

Art nouveau glass and metal entrance to Porte Dauphine metro station, Paris, France, designed by architect Hector Guimard ca. 1900. The Art Archive/Manuel Cohen.

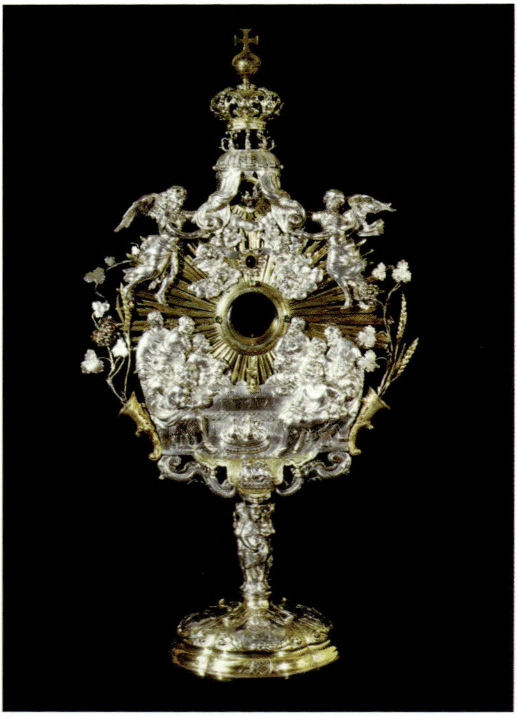

Baroque monstrance or communion vessel by Johann Zeckl, 1705. Silver with partial gilding. Height 33.3 inches (84.6 cm). The Art Archive/Victoria and Albert Museum London/Eileen Tweedy.

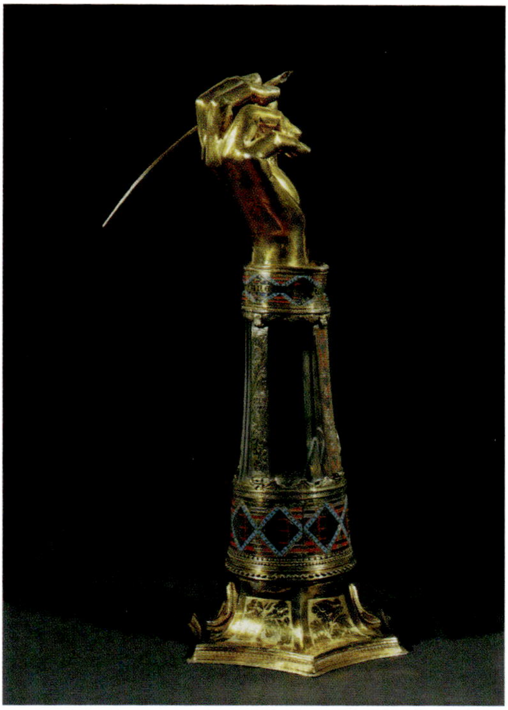

Reliquary, arm of Saint Luke, Naples, ca. 1336–1338. Gilded silver, enamel, rock crystal. From the Treasury of Medina del Campo. Erich Lessing/Art Resource, NY.

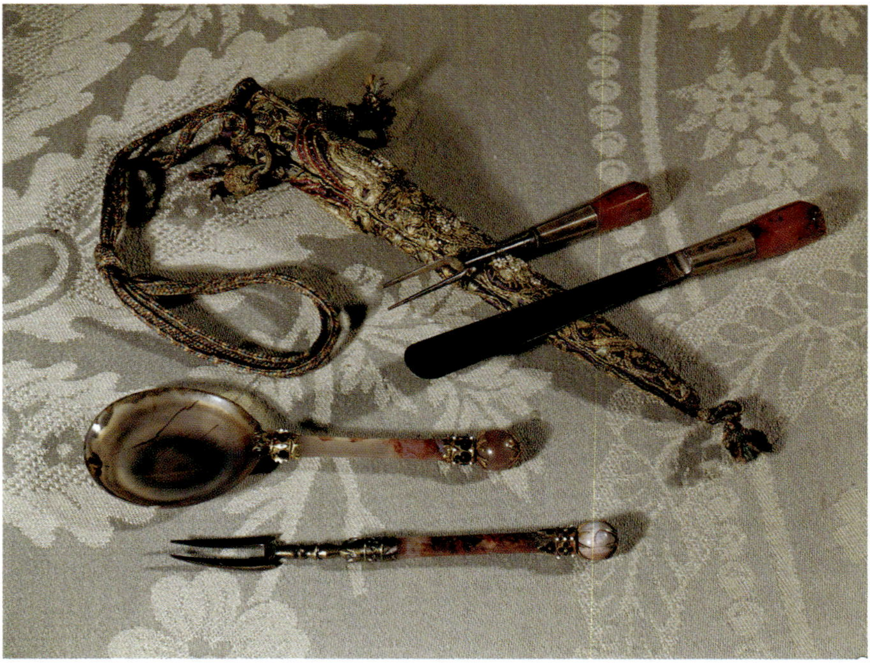

Travel set of flatware and bag of Queen Elizabeth I (1533–1603). Location: Collection of the Duke of Northumberland, Alnwick Castle, Northumberland, Great Britain. Erich Lessing/Art Resource, NY.

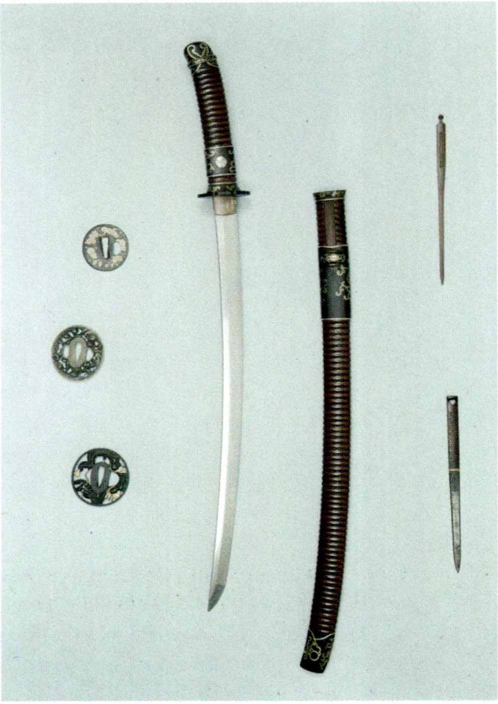

From left to right: three examples of Japanese *tsuba,* a katana with sheath, a sword needle (*kôgai*), and a sword knife (*kogatana*) that would have hung from the *tsuba* holes. The Art Archive/Gunshots.

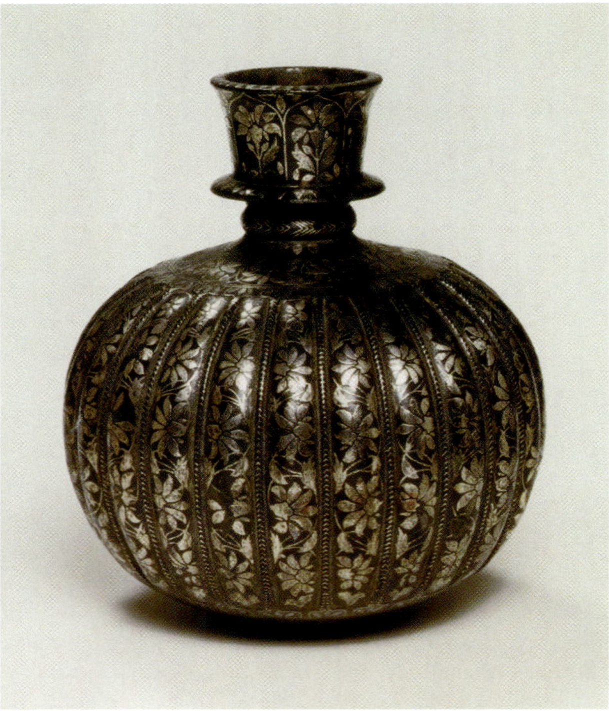

Huqqa bowl cast in the *bidri* alloy with silver inlay and patina. Late seventeenth to early eighteenth century. Height: 6.5 inches (16.5 cm). Location: Victoria and Albert Museum, London, Great Britain. Victoria & Albert Museum, London/Art Resource, NY.

made of interlocking pieces that may be removed from a vessel with a reduced diameter. If a chuck is solid and in one piece, it must not reduce in diameter from the center, as undercuts in the spun forms would make it impossible to remove the finished object

Almost any metal may be formed through spinning. The rigidity of the metal and the depth of the draw determine how often it must be **annealed.** Excessive pressure on the metal being spun may result in its being pinched between the tool and the chuck, which may cause thinning and tearing. Lubrication is added to the metal being spun to reduce friction. Rough edges may be trimmed while on the lathe by the use of a cutting tool.

Although there is some evidence to suggest that spinning may have been used as early as **ancient Roman** times, it was not until the use of steam power that the process became truly effective and adopted by metalworkers on a large scale. Spinning has been used in quantity production since the first half of the nineteenth century. The advent of greater power driving capacity made the spinning lathe an important element of the industrial revolution, and it is still used as a form of mass production today. Spinning is not as adaptive a process as **raising** for the making of hollowware, and so is best used when many identical forms are desired. Commonly spun utilitarian forms today include aluminum and stainless steel plates, bowls, and cookware.

See also Art Deco; Bauhaus; Modernism; Nineteenth Century; Stamping and Die Forming

FURTHER READING

Deutsches Museum, Munich, Germany, http://www.deutsches-museum.de

Untracht, Oppi. *Metal Techniques for Craftsmen.* Garden City, NY: Doubleday, 1968.

Woodbury, Robert S. *Studies in the History of Machine Tools.* Cambridge, MA: M.I.T. Press 1972.

Lead. ***See*** Metals

Liturgical Metalwork. ***See*** Middle Ages

Metals

A metal is a member of a class of chemical elements or one of those elements in combination with something else. A metal that is not a pure element is known as an alloy. Alloys may be a combination of two or more metallic elements or a combination of a metallic element with a nonmetallic substance such as carbon.

Metals usually occur in the earth's crust in combination with other substances. This mixture is known as an ore. When they are mined, ores must be processed to separate out the individual metals (see **Mining and Smelting**). Copper, silver, gold, and platinum also occur in surface deposits in relatively pure states, known as native metals. This makes them more likely to have been used by early peoples (see **First Metalsmiths**).

Metals are usually characterized by the following qualities: ductility, malleability, luster, density, and (with the exception of mercury) solid state at room temperature. In addition, metals are generally capable of conducting heat and electricity. The two main ways of shaping metal are by melting and **casting** it or by forming it through the application of force.

In general terms, metals are commonly described as being either ferrous (iron-bearing) or nonferrous (not iron-bearing). The nonferrous metals can be further divided into noble metals and base metals. Noble metals are elements that are resistant to tarnish and are more chemically stable. Noble metals also tend to be considered precious. In general, base metals are found in greater abundance and are usually more chemically reactive than noble metals.

NONFERROUS METALS

Aluminum (Element, Base)

Aluminum is a silver-white metal that is malleable and ductile, has a low density and a low melting point, and resists corrosion. It conducts heat and electricity well and is the most abundant metallic element on the planet, comprising eight percent of the earth's crust.

Aluminum also combines well with other materials to form alloys that are stronger but still light. Some of the materials that aluminum may be alloyed with include magnesium, silicon, copper, manganese, titanium, silver, and zinc. Aluminum is used contemporarily in the manufacture of cooking vessels, automobile and aircraft parts, and scientific instruments. Aluminum may be colored through the process of **anodizing.**

Aluminum was not isolated until 1825 and could not be produced in significant quantities until 1854. An inexpensive method of extraction was finally developed in 1889. In the interim, aluminum was so costly that Napoleon III had his best set of flatware made from it.

Copper (Element, Base)

Copper is an orange/pink metal that is used in many alloys. It is malleable, ductile, and capable of conducting heat and electricity very effectively. Because copper reacts well with oxygen, copper and its alloys take **patinas** very well. Its reactivity to oxygen also means that copper does not cast well in its pure form.

Because it conducts heat well and is relatively inexpensive, copper has been used extensively in food preparation. Due to the fact that copper reacts with air and water to form poisonous substances on its surface, it must be cleaned before each use and never used with acidic foods. For these reasons, copper cookware is frequently coated with tin or some other less reactive metal.

Copper may have been worked as far back as 13,000 B.C. and was probably the first metal used by humans. The word "copper" comes from the Greek name for the island of Cyprus, which is where the Romans got their copper.

Gold (Element, Noble)

Gold is heavy and yellow in color and was the first of the noble metals to be used by humans. It conducts heat and electricity very well and is highly malleable and ductile. Gold casts well and may be drawn into very fine wire or pounded into sheets so thin that it becomes translucent. Some gold objects from Egypt date to before 3000 B.C.

Gold is frequently alloyed to produce a less expensive material or to change its color or working properties. Contemporarily, gold quantities are expressed in karats, as ratios of parts per 24. Pure gold is expressed as 24 karat. Eighteen karat gold is 18/24 or 75 percent gold. Lower karat gold alloys are usually made by adding some quantity of copper and silver to gold. White gold is a gold-based alloy that is designed to appear light gray or white in color. Many recipes exist, but the two most common metals alloyed with gold to whiten it are nickel and palladium (a member of the platinum family). Zinc, silver, and copper are also frequently included in such mixtures in order to improve the working characteristics of the alloy.

Lead (Element, Base)

Lead is a heavy, dense metal of a blue-grey color. It is not very elastic, but its low melting temperature makes it easy to cast. Lead also alloys well and is corrosion resistant. Records of this metal date back to 3000 B.C. In ancient times, it was used in plumbing, the decorative arts, and in architecture.

Nickel (Element, Base)

Pure nickel is hard, malleable, ductile, and light gray in color. It resists oxidation and is variably magnetic. The first samples were found in meteors, but it was not isolated as an element in the earth's crust until 1751. Nickel alloys well and is used as plating to prevent corrosion. Alloys of copper, nickel, and zinc were used in China in the Middle Ages. Nickel is now used extensively by industry, frequently for electroplating (see **Gilding and Plating**).

Nickel silver is a misnomer insofar as there is no silver in this alloy. The term is used to describe an alloy that is mostly copper with the addition of nickel (and sometimes other base metals) to produce an alloy that resembles silver in color. German silver has a makeup similar to nickel silver, but with additional zinc that gives it a light yellow hue.

Platinum (Element, Noble)

Platinum is a white-gray metal. It is dense, corrosion resistant, ductile, malleable, and has a high melting point. Platinum is the most frequently occurring of the platinum group of metals, which also includes: palladium, rhodium, iridium, ruthenium, and osmium. Contemporarily, platinum is used by the chemical industry and in jewelry and dentistry. Rhodium is sometimes used as a thin plated coating over silver objects in order to improve their whiteness.

Platinum may be found as relatively pure nuggets in riverbeds in certain parts of the world. A few isolated, early objects made from platinum have been found from as early as the seventh century B.C. Prior to the nineteenth century, there was no technology available with which to reach the high melting temperatures of the platinum group. Objects that pre-date that technology are comparatively small pieces that have been simply **forged** and soldered (see **Connections**).

Silver (Element, Noble)

Silver is a lustrous, light gray metal that is very ductile and malleable. Pure silver is also sometimes referred to as fine silver. The earliest silver objects made date to around 4000 B.C. Today most of the world's silver is consumed by photographic and x-ray processes.

Most metalworking techniques require that pure silver be alloyed with copper, which adds strength without causing too much change in its lustrous white appearance. A common alloy, known as sterling silver, is 92.5 percent silver and 7.5 percent copper. This may be indicated on a more contemporary object with the inscription ".925." The recipe for sterling silver dates back to the twelfth century when King Henry II adopted the use of a silver alloy from Easterling, Germany.

The term coin silver can be misleading. In times when coins were made of precious metals, whatever alloy was being used in the region at the time was their coin silver. This varied by locality.

Tin (Element, Base)

Tin is a blue-white metal. It is malleable at room temperature but becomes brittle when heated and casts well. It resists corrosion and thus is frequently used to coat more reactive metals that are going to be in contact with food. The term tinplate is used to describe objects that are made of iron or steel and coated with pure tin. It may also be found as a component of bronze, pewter, and Britannia.

Titanium (Element, Base)

Titanium is a light gray metal with a low density and high melting temperature. It is also strong and corrosion resistant. Because of these characteristics, it was named for the titans of Greek mythology. Its high reactivity to oxygen makes it difficult to cast and fabricate with. It is commonly alloyed with aluminum, tin, nickel, palladium, and vanadium in order to change the working properties.

Titanium as an element was not discovered until 1791 and is the ninth most abundant metal in the earth's crust. It was not until the twentieth century that a commercially viable way of processing it was developed. Contemporarily, it is commonly alloyed with steel for aircraft components and weapons. Because it is strong and nonreactive, titanium is also used for surgical implants and chemical processing. Titanium may be colored through **anodizing.**

Brass (Alloy)

Brass is typically an alloy of copper with zinc in a two-to-one ratio. This produces a bright yellow metal that is hard, strong, and more resistant to corrosion than copper. Brass cuts cleanly and can be somewhat malleable depending on the recipe. Some brasses contain lead. Many different recipes for brass have been used historically, and in some older writings the terms brass and bronze seem to be used without differentiation. Brass has been used extensively for domestic objects and hardware.

Pinchbeck is an alloy of copper with a small amount of zinc that was invented by Christopher Pinchbeck of London in the seventeenth century. It was used extensively during the Victorian period in jewelry applications where a cheaper imitation of gold was desired.

Britannia (Alloy)

Britannia is a light gray alloy of mostly tin with copper and antimony. It is hard and somewhat malleable and may be used in both fabrication and casting processes. Britannia was developed in England in the nineteenth century as a harder, more silvery and lead-free alternative to pewter. It has been largely used for domestic objects.

Bronze (Alloy)

Bronze is the result of alloying copper with tin. The addition of a relatively small percentage of tin (5–15%) results in a much harder metal that has a lower melting temperature and casts well. Bronze might also include other metals in the recipe, such as manganese, lead, and iron. Traditional bronze recipes produce a metal with a golden-brown color. Because of the high copper content, bronze also takes patinas quite well.

The term bronze has also been used as a general descriptor for any alloy of copper with a small amount of other metal. For example, a mixture of 96 percent copper and 4 percent aluminum is referred to as aluminum bronze. Gunmetal is a copper alloy containing large amounts of tin and zinc. Speculum metal is a bronze with a high tin content and sometimes lead. Speculum metal has been used for mirrors and is sometimes called white bronze because of its lightness of color.

Electrum (Alloy)

Electrum is a naturally occurring alloy of gold with 20 to 40 percent silver. It is pale yellow in color and was used in ancient times for coins and decorative objects.

Paktong (Alloy)

Paktong is a strong, light gray alloy of nickel, copper, zinc, and iron. It was produced in China as early as the sixteenth century for use in domestic objects and coinage. By the late seventeenth century, *paktong* was imported to Europe, where it was also referred to as white copper.

Pewter (Alloy)

Pewter is made of tin alloyed with smaller percentages of other metals, which may include copper, lead, zinc, bismuth, and antimony. Pewter is malleable and casts well. Its low melting temperature means that it may be cast into a wide range of mold materials. A pewter flask from around 1450 B.C. was found in Egypt, and pewter has been in common use for household goods since antiquity.

Sentoku (Alloy)

Sentoku is a **Japanese** alloy composed of copper, tin, lead, and zinc. It is most often used in pictorial compositions for its yellow color. It is frequently juxtaposed with *shakudo* and *shibuichi* in *tsuba,* Japanese sword guards.

Shakudo (Alloy)

Shakudo is a Japanese alloy composed mostly of copper with three to five percent gold. Sometimes a very small percentage of silver is also used. It is generally used because of the blue-black color it takes on when patinated. It is frequently found alongside *sentoku* and *shibuichi* in *tsuba,* Japanese sword guards.

Shibuichi (Alloy)

Shibuichi is a Japanese alloy made up of approximately 75 percent copper and the remainder in silver. It is generally used for its light grey color alongside *shakudo* and *sentoku* in *tsuba,* Japanese sword guards.

White Metal (Alloy)

White metal is the name for an alloy of tin and antimony. It has a light gray color and low melting temperature. It is used to produce inexpensive jewelry that somewhat resembles silver.

FERROUS METALS

Iron (Element, Base)

Iron is a dark grey metal that is hard, strong, ductile, malleable, and magnetic. It is the most common of metals, and it is believed that the earth's core is mostly iron. Humans have known meteoric iron for nearly 5,000 years. We have known about it from surface deposits for half as long. Iron oxidizes easily, which results in rust (oxidization is the chemical reaction of a metal with oxygen, which combine to form a metallic oxide).

Pig iron is the result of casting unrefined, smelted iron into oblong forms known as pigs (smelting is the melting of an ore to separate its metallic contents). Pig iron itself has a three to four percent carbon content. That crude iron, or pig iron, is then used to make cast iron, wrought iron, or steel. Pig iron appeared in China around the tenth century B.C. It appeared in Europe around the fourteenth century A.D.

Cast iron is a dark grey metal containing carbon and silicon. It is strong, dense, hard, and brittle but has a low melting temperature and is good for casting. It is made by heating pig iron until the carbon level is down to two or four percent and most contaminants are removed. It is then cast into its final form. It is used today in engine blocks for the automobile industry and construction materials.

Wrought iron contains less than .3 percent carbon and is malleable, ductile, tough, more corrosion resistant than regular iron, and good for **forging.** This is commonly used for architectural work such as bridges or gates. To make steel, this is further refined and alloyed according to its intended use.

Steel (Alloy)

Steel is a dark gray, hard, flexible metal. It is an alloy of iron containing less than 1.7 percent carbon as well as other metals. These may include nickel, titanium, chromium, manganese, tungsten, and more. The hardness of steel depends on its carbon content. Steel has been in use since around 1400 B.C. and is most commonly used for tools, weapons, and architectural structures.

Stainless steel is a general term for a group of steel-based alloys that may include chromium, nickel, silicon, and others with an iron base. It is characterized by its toughness and resistance to corrosion. Stainless steel was developed in Europe early in the twentieth century.

FURTHER READING

Aitchison, Leslie. *A History of Metals.* New York: Interscience Publishers, 1960.

Betteridge, W. *Nickel and Its Alloys.* Chichester, West Sussex, England: Ellis Horwood, 1984.

Fleming, John, and Hugh Honour. *The Penguin Dictionary of Decorative Arts.* London: Viking, 1989.

Haedeke, Hanns-Ulrich. *Metalwork.* Vivienne Menkes, trans. New York: Universe Books, 1970.

Lütjering, G., and J. C. Williams. *Titanium.* Berlin: Springer, 2003.
Mantell, Charles Letnam. *Tin: Its Mining, Production, Technology, and Applications,* reprint. New York: Hafner Publishing Company, 1970.
U.S. Geological Survey, http://minerals.usgs.gov/minerals

Middle Ages
(ca. 350–1450)

The period between the fall of the Roman Empire and the **Renaissance** in Western Europe is generally referred to as the **Middle Ages,** the medieval period or by the more negative term, Dark Ages. It also encompasses the stylistic periods of Carolingian (eighth to tenth century), Romanesque (ninth through twelfth century), and Gothic (twelfth through fifteenth century). The formation of the countries of Western Europe as we know them today began during the Middle Ages.

During the fourth century, Huns displaced a number of Germanic tribes who were forced into the Western Roman Empire. The various tribes settled in North Africa, Italy, Spain, France, and Great Britain. As the Roman Empire broke down, the West divided into a number of insular kingdoms that were a mix of Germanic and Roman cultures. The far-reaching trade of the Roman Empire came to an end, and local government began to revolve around wealthy landowners. The use of coinage declined, as did the presence of a middle class. As town populations shrank or disappeared altogether, people were driven into religious communities or to work the land of the wealthy. Society was essentially reduced to the wealthy landed nobility, the increasingly powerful church, and the peasantry.

During the Middle Ages, Christianity spread, and most of Western Europe converted to Roman Catholicism. The arts of the Middle Ages were largely devoted to religious ritual and the greater glorification of the church. During a period of peace within Western Europe, from the eleventh through thirteenth centuries, considerable strides were made in improved technologies, the economy, and general quality of life. It was at this point that there was a reappearance of merchants and a resurgence of towns. Craftsmen started to move out of the church and away from the control of the nobility. The **guild** system began to develop, and towns grew into self-governing entities.

GOLD AND SILVER

The landowners, aristocracy, and powerful members of the church covered their tables with silver and gold, both for use and for the display of wealth and status. Silver and gold tablewares were displayed as tiered arrangements on sideboards and buffets. By custom, the number of shelves one displayed was determined by rank. In addition to the ownership of grand objects, it also became customary for members of the nobility to exchange gifts of silver and gold items with one another.

Objects that were made in gold and silver included dishes, jugs, basins, spoons, beakers, chalices, and bowls. Through the eighth century, decorative imagery included abstracted figural and animal motifs as well as the insignias of patrons. Representations were created with **engraving, niello** inlay, the setting of precious stones, **enamel,** and **chasing and repoussé.** Until the thirteenth century, some cups incorporated carved stone or coconut shells mounted as the bowl. During the Carolingian dynasty, stylistic influences came from Italy and the rest of the Mediterranean. Imagery began to recapture classical tendencies as well as the influence from the **Byzantine Empire.** Throughout the Romanesque and Gothic periods, ornament in metal followed that of architecture.

Members of the nobility frequently employed a number of craftsmen on their lands to produce the goods for their tables and **armor and arms** for their military needs. These were more likely to be generalist metalworkers who were able to both make plate for the table and to fix household wares. A few specialists traveled the countryside for commissions from either churches or noblemen, such as **Nicholas of Verdun**.

Monasteries became more important after the fall of the Roman Empire, and even metalworking moved into the religious communities. Eventually the church began to separate members of their community who were strictly monks from lay brothers, those who had taken vows but were at the monastery primarily to practice a craft.

In general, education had declined in the Middle Ages, and most of the poorer classes could not read. This made the visual symbolism in churches even more important as a way of communicating the messages of the Bible. The church used displays of gold and silver to project an image of splendor as well. Precious materials were used to impress the significance of the objects and their associated rituals upon the devout.

Each church had a set of liturgical plate. The liturgy is the celebration of the Christian Communion, in which wine and bread are blessed and given out. Roman Catholics believe that these become the blood and body of Christ upon consecration. The paten or plate is a shallow dish used to carry the consecrated bread. Also needed are the chalice for the consecrated wine, a container for the bread, and a vessel for the wine. Other metalwork that a church might own could include covers for books, crosses for carrying or display, censers for burning incense, candle holders, lanterns, altars, and reliquaries.

Reliquaries were generally the most costly and elaborate metalwork commissioned by a church. A reliquary is a container for a relic, usually the remains or former possession of a saint or some other significant person. The most grand were over six feet long and covered with gold, enamel, and precious gems. The understructure was often wood built into a boxlike form. From the mid-eleventh century through the early thirteenth century, such shrines were shaped like coffins with recessed areas containing sculptural figures. Following this, the trend was toward more architectural shapes, usually resembling the forms of period churches. Exterior imagery on these boxlike reliquaries was derived from the Bible and the life of the person from whom the relic came. Figural reliquaries were shaped to mimic the body part from which the relic of the person had come. For example, a foot bone might be enshrined in a reliquary made in precious metals and gemstones to resemble a foot.

The imagery on church objects was often executed in enamels. Additional ornamentation might include gems, chasing and repoussé, **cast** elements, **stamped** motifs, engraving, and niello inlay. Aside from these large public pieces, goldsmiths also made smaller images for the private devotion of the affluent.

HOLY GOLDSMITHS

The Roman Catholic Church establishes which deceased persons are to be considered saints through canonization. Patron saints are specifically affiliated with a place or group of people. It is believed that the patron saint guards over the best interests of those whom it represents and that they may be prayed to for intercession. A saint's patronization is often based on events from his time on earth or a perceived relationship with the location or group of people. This may be determined by the church or by the people who choose to pray to such saints. For this reason, there are overlapping patronages in many professions.

Several saints have been associated with the metalworking trades. Saint Eligius was a metalsmith at a mint during his lifetime in the late sixth and early seventh centuries and is thus considered a patron saint of metalworkers. Saint Dunstan of Canterbury is considered a patron saint of goldsmiths because he had been a practicing goldsmith during the tenth century. He is also sometimes affiliated with blacksmiths and locksmiths. Saints Barbara, Anastasius, Brigid of Ireland, and others have also been linked to the metalworking professions.

As a rule, the style of church metalwork changed very slowly. During the early Middle Ages it largely continued the formalism of the late Roman Empire. Over time both the naturalism of the figures and their visual symbolism increased. In general, Germany led stylistically until the twelfth century, after which France and then Italy became the principal originators of style. In both houses and churches, when silver and gold could not be afforded, the same wares were made in bronze, brass, copper, or pewter.

BRONZE, BRASS, AND COPPER

During the Middle Ages, most bronze casting was done for churches. Groups of bronze casters traveled from one major building project to the next. A new church might require their services in order to make bells, doors, fountains, lecterns, candelabra, statues of saints, tombstones, and crucifixes. As towns and guilds developed in later years, these itinerant groups settled into centers of production from which the finished work was exported.

After the eleventh century, Western Europeans who had participated in the crusades began bringing back goods from the Middle East. Some of the metal objects became models for Western goldsmiths, such as the aquamanile. These are sculptural pitchers for hand washing, both domestically and in religious ceremony, and are usually shaped like animals or heroic individuals. These were cast with the lost wax process.

In the second half of the Middle Ages, around the twelfth century, brass became more widely used. It was both cast and **raised** to form dishes, basins, fire irons, and candlesticks. A significant industry in the production of bronze and brass wares was established in the town of Dinant in what is now Belgium. The town's name was so synonymous with the craft of metalworking that bronze and brass objects were sometimes referred to as dinanderie, regardless of their place of origin. What they were best known for were shallow basins with stamped or chased decorations in the center.

Copper was another, less expensive substitution for silver and gold at church or on the table. In order to be used for a Communion chalice copper had to be **gilded** on the interior as not to react with the wine. Gilded copper was also used in place of gold for enameled reliquaries and other church equipage.

PEWTER

In the Middle Ages the use of pewter became more widespread. Pewter made a good substitution for silver in the Communion ceremony because it did not interact with the wine or food. It was also used over wood to form containers for consecrated wafers. Pewter was even decorated to simulate gold **filigree,** although designs had to be cast in pewter, so they were not true filigree.

A vessel form called an ampulla was frequently made in pewter for wide distribution. Ampullae have rounded bodies and narrow necks and were originally used by Romans to hold oil or perfume. The Christian church adopted the form to hold consecrated oil. Pilgrims bought pewter ampullae as souvenirs of a pilgrimage. These were often illustrated with scenes from the Bible and were popular throughout the Middle Ages.

From the late Gothic into the Renaissance period, pewterers around central Europe made some of their most imaginative work for the other guilds. Guild flagons, common to the areas that are today Poland, the Czech Republic, and Germany, were a part of the guild's events and celebrations. These were generally very large and sometimes as tall as 30 inches. At guild functions, members filled their own tankards from these large flagons. Some were so heavy when filled that they were equipped with a spigot at the bottom. The flagons bore the guild's name and usually some pictorial reference to their occupation. There were regional distinctions in the general forms of flagons and other pewter tableware. As the general prosperity increased, pewter replaced the wooden dishes of the lower classes.

IRON

Toward the end of the Middle Ages, wrought iron was increasingly used for its strength and ability to reinforce wood. Iron hinges on church doors kept the vertical wooden planks together and helped to keep the gold and silver valuables within the church safe. During the eleventh through thirteenth centuries, these hinges became more elaborate. The height of Middle Age ironworking may well have been the kind of elaborate Gothic door hardware made to resemble scrolling foliage that can be seen throughout Europe.

Iron also began showing up inside the church in the form of screens and ornamental railings. The ironworkers occasionally branched out to make freestanding objects for church interiors such as chandeliers and candle stands. Closely related to the door hinges were the iron bands made to reinforce wooden chests where valuables were kept. Iron was not used for domestic door hardware until the late fifteenth century.

ENAMEL

The Middle Ages was an important period in the history of enameling. A number of new techniques were developed, and some important centers of production were formed. Enamels were also a striking way to create colorful symbols with which to communicate to the public through objects. Much of Western European enameling was inspired by Byzantine work.

Like other areas of metalworking, enameling took refuge in the monasteries and was practiced there until public demand increased. Towns such as Limoges in France became famous for their enamel work. As larger pieces were needed, enamelists used gilded copper instead of gold to make the products more affordable.

Cloisonné was used throughout the Middle Ages but its popularity declined with the increased use of champlevé in the twelfth century (see **Enamel**). Champlevé was first done with colored figures set against a gold or gilded background, but in the late twelfth century these became inverted. In the late thirteenth century, the enameling technique of *basse taille* was invented in Italy.

In the late fourteenth century a technique for enameling objects in the round was developed. This was called *en ronde bosse,* in rounded relief. It was used with great success on figures in reliquaries. Also toward the end of the Middle Ages, *plique-á-jour* was developed in France. The technical proficiency needed to create these suspended enamels demonstrates the degree to which enamelists were perfecting their medium. Two additional techniques, painted enamels and *en grisaille*, brought an increased delicacy in the handling of enamels. Painted enamels (enamel applied in a painterly manner) were first used in Flanders, and the effect resembled miniature paintings. *En grisaille* enameling involves the slow buildup of layers of white enamel on a black background. The varying levels of opacity produced a range of grey, allowing for a modeled effect.

See also Arts and Crafts Movement; Celtic People; Flatware

FURTHER READING

Belli Barsali, Isa. *Medieval Goldsmith's Work.* London: Hamlyn, 1969.
The British Museum, http://www.britishmuseum.org
Cherry, John. *Goldsmiths. Medieval Craftsmen.* Toronto: University of Toronto Press, 1992.
Coatsworth, Elizabeth, and Michael Pinder. *The Art of the Anglo-Saxon Goldsmith: Fine Metalwork in Anglo-Saxon England: Its Practice and Practitioners.* Woodbridge, England: Boydell Press, 2002.
Haedeke, Hanns Ulrich. *Metalwork.* New York: Universe Books, 1970.
The Metropolitan Museum of Art, the Cloisters Curatorial Department, http://www.metmuseum.org/Works_of_Art/the_cloisters
National Gallery of Art: Virtual Tour of Medieval Metalwork and Enamels, http://www.nga.gov/collection/gallery/medieval/medieval-main1.html
Notre Dame Cathedral, Paris, http://www.notredamedeparis.fr
Victoria and Albert Museum, William Walter Watts, and John Starkie Gardner. *Ironwork.* London: Board of Education, 1927.

Mining and Smelting

Mining is the process of purposely removing a material, such as metal, from the earth. Metal mining began in some areas as early as the fourth millennium B.C. Smelting is the melting of an ore with the intention of producing metal from it. It is believed that smelting furnaces developed out of the technology of firing pottery. An ore is a material bearing one or more minerals. Refining is the process of increasing the purity of a metal by removing traces of other metals. A flux combines with impurities in molten metal, allowing for their separation. Fluxing agents reduce the amount of metal lost in slag formation.

The history of the acquisition and processing of metals is closely tied to the development of civilization itself. Today's mining process rely largely on machine labor, but at one point mining was the work of thousands of slaves. The contemporary use of toxic chemicals, displacement of large areas of earth, waste created by mining processes, and effects on local populations caused by mining today are subjects of concern for many. The following are brief descriptions of how some of the major metals have been mined and prepared for use over the course of history.

GOLD

The first gold used by humans came from alluvial or so-called placer deposits found in riverbeds. These are caused by natural erosion of the metal-bearing minerals on the earth's surface. Placer deposits are mined by sifting through the deposited materials, allowing the heavier gold to separate itself by sinking. In early societies, cloth or animal hides were sometimes used to catch the small particles of gold as they were washed over by the water. The use of sheepskin in such a manner may be the basis of the Greek myth of the Golden Fleece. Later, individual miners used pans to sift small amounts of surface deposits. In Africa, the ground gold deposits were washed in containers made from gourds. Today, placer deposits are mechanically sifted in large containers.

By the third millennium B.C. people had begun to dig into the earth in search of gold. The metal could be found alone or in combination with silver running through minerals such as quartz. The metal and surrounding rock were cut from the earth and broken up by pounding and grinding them between harder stones. Larger pieces were separated out and the remaining dust and small particles were washed across sloping tables, where the metal's differentiation in weight caused a physical separation from the rock.

By the end of the first millennium B.C., mercury amalgamation was being used to separate metals from other minerals. Gold-bearing minerals were heated with mercury, into which the metals present in the rock dissolved. The mercury was then distilled off, and the gold and silver remained. A similar amalgamation process was also used with lead instead of mercury. The use of cyanide to leach metal from ores was discovered in the late nineteenth century. Subsequent discoveries have continued to improve the refining process.

Because gold and silver were frequently found intermingled, methods for separating the two were developed. One such process used in Mesopotamia from the second millennium B.C. involved mixing the fine metal particles with salt. Silver formed a chloride with the salt that could then be removed with water.

Contemporarily, rock bearing gold and other metals is removed with drills and explosives. The two major kinds of mines are open pit and underground mining. In open pit mining, the metal being sought is located near the surface of the earth and can be reached by systematically digging a large, stepped hole that is entirely exposed. In underground mining, the metal sought is usually a considerable distance beneath the surface. Underground mining involves digging tunnels deep into the earth to extract the ore.

Today different methods are used to separate gold from the ores it is found within. It can take 10 to 30 tons of ore to produce one troy ounce of gold. After the ore is removed from the earth, it is crushed and ground by machines. Gold is then leached from the powdered rock, usually using cyanide. Gold and other metals may be removed from this solution in a variety of ways, including electrowinning, which uses the same principals of electrolysis as **electroforming,** and carbon-in-pulp, which uses carbon particles, cyanide, and sodium hydroxide.

In the carbon-in-pulp process the cyanide is introduced at the ore grinding stage. Oxygen is added to help dissolve the gold, which is then absorbed by small carbon particles. The particles are removed from solution and the gold and other metals are removed with a solution of cyanide and sodium hydroxide

Heap leach mining is used when the ore contains a very small percentage of gold. Cyanide solution is filtered through hills of extracted ore and collected from its base to be refined. Although the mercury amalgam process is still used, the use of cyanide to extract gold from its ores is most common now. The Miller purification process uses gaseous chlorine, which combines with most of the nongold element. The Wohlwill process utilizes electrolysis.

SILVER

The earliest silver used was found in surface deposits in the same manner as gold. Silver is primarily found in three ores: red silver ore, horn ore, and argentite. Argentite is the most common of these, being a combination of silver and sulfur. Silver may also be a byproduct of the mining of other metals, such as lead, gold, or copper. It is from lead and copper ores that most silver is extracted.

By the second millennium B.C. people were mining lead ores in order to smelt out the silver, which was later purified through cupellation. Cupellation is the process of exposing metal ores to high temperatures and blasts of air, which oxidizes the metals of lower melting temperatures, such as lead, making them easier to separate out.

In South and Central America, Spanish explorers mixed ground silver ores with heated copper ores, salt, and mercury. The grinding was done by mules walking over paving stones covered with the ingredients. Thus this process got the name of the patio process. Over time the silver broke down and bonded with the mercury, which was then removed and melted away, leaving the silver. At the end of the nineteenth century the use of cyanide replaced the patio process.

Today, silver is obtained by breaking up the ores and then smelting it. In the case of copper ores, the mixed metal alloy that results is refined to produce the copper, leaving a silver residue called sludge. Nitric acid is used to dissolve the silver from the sludge. In order to precipitate the silver from this solution, electricity is used. When a lead ore is processed for silver, zinc is added during smelting, which bonds with the silver content and floats to the surface. Once separated, the zinc is separated from the silver with heat treatments.

COPPER

Several groups including the **ancient Egyptians** mined copper in ancient times. The process of extracting copper has changed very little since then.

Historically, broken-up copper ore was smelted, which separated it from impurities but not other metals. This crude copper was then reheated and mixed with silicic acid found in some clays. Repeated melting continued the refining process. This mixture was then stirred with a length of young, green wood coated with coal dust, which would have further helped to remove oxides and impurities.

In contemporary copper mining, the crushed copper ore is concentrated somewhat prior to smelting. This is done through a process called flotation. In flotation, the ore particles are put in a tank with chemicals that attract the metal-rich particles and cause them to float to the top, where they can be skimmed off. After smelting, the copper goes into a converter, which keeps the copper melted while blasts of air and silica are added. The silica acts as a fluxing agent, drawing off impurities. The final stage of the copper refining is melting the copper in a reducing or fuel-rich environment using natural gas, which consumes oxygen and other unwanted gasses. Copper may also be extracted from its ores through leaching. Although both of these processes render a copper product that is 99 percent pure, even greater refinement may be achieved though the use of electrolysis.

TIN

Tin is available in alluvial deposits but rarely found in pure form. The most common tin ore is cassiterite. The smelting of tin was vital to the development of the Bronze Age and later to the making of pewter.

In contemporary mining practices, tin ores are heated to remove the sulfur content. This is sometimes used in conjunction with leaching, to remove more impurities. The tin and its remaining impurities are smelted with carbon and limestone, which acts as a flux. The resulting tin is further refined. Refining may be done by melting the tin and blowing in air, which bonds with the impurities and floats to the surface, where they may be skimmed off. Tin may also be melted out of impurities with higher melting points. Further refining may be done with electrolysis.

ZINC

Zinc does not occur alone in nature, only in combination with other materials. Until around 1800, brass was made by smelting copper with calamine, an ore containing zinc oxide. The knowledge of zinc as a separate element and how to extract it is relatively recent. There is evidence of zinc smelting in India in the fifteenth century, but it was not until the seventeenth century that it was being accomplished in the West. When heated, zinc within the ore becomes zinc oxide. This can be converted to zinc though the inclusion of carbon, but this reaction actually takes place above zinc's boiling point. This means that the gaseous zinc has to be caught and condensed.

Increasingly efficient methods of processing zinc have been developed in the eighteenth and nineteenth centuries. The contemporary process involves crushing and grinding the rock that bears zinc ores and putting it in a tank with chemicals that cause the zinc-bearing ore to rise to the top, where it can be skimmed off. This more concentrated material may be processed in a number of ways, the most common of which utilizes electricity. An electric current is passed through the solution containing zinc. The anode, or positively charged electrode, is made of lead and silver, and the cathode, or negatively charged electrode, is made of aluminum. The current forces the zinc out of solution and onto the aluminum cathode.

IRON AND STEEL

The oldest form of iron production involves heating iron ore and charcoal in a tall furnace or cupola. The high temperatures needed were achieved by blowing air into the fire. Impurities known as slag were periodically allowed to run out of a hole at the bottom of the cupola. When the time was right, a soft form of iron would be removed from the furnace (see **Africa**). This spongy iron needed to be reheated and forged to create useful implements. This was the dominant method of producing iron through the Middle Ages.

In the fifteenth century, water power was harnessed to force greater amounts of air into the burning fuel, thereby reaching higher temperatures within the furnace. The iron actually became molten and was permitted to pour out of the cupola when it was ready, resulting in cast iron. Additional smelting produced a fairly pure form of iron known as wrought iron. In the eighteenth century, the charcoal fuel was replaced by coke, a distilled form of coal.

Today iron ore is usually dug from open pit mines and then crushed and ground by machines. The iron-rich particles are separated out by their heavier weight or magnetic attraction, depending on the kind of ore being processed.

The two major means of smelting iron contemporarily are the blast furnace and the direct reduction furnace. A blast furnace works in a similar manner to the old cupola method. The furnace is a tall cylinder into which raw materials of ore, coke, and limestone, a fluxing agent, are fed from the top. Molten iron is recovered from an opening near the bottom. The impurities that separate out of the iron are collected intermittently at a point just above the pool of molten iron. A blast of hot air is forced through the materials from the base of the furnace, which helps the coke to burn at a very high temperature and expel any oxygen present. This process continues day and night for years, with more raw materials continuously being added until the furnace itself requires repair.

In direct reduction furnaces, the iron ore is exposed to gasses, such as natural gas, which consume or burn oxygen. A slag of impurities is not created in this process; therefore, the iron from a reduction furnace must undergo additional purification processes.

In turning iron into steel, the primary change is the reduction of the carbon content. The development of a reliable way to make steel had a significant effect on the course of history. The way steel was initially made was to heat iron in the presence of carbon in the form of charcoal. This had to be done in a reducing or oxygen-free environment so that the iron would absorb some of the carbon (see **Lamination Techniques**). Many specialty steel alloys exist that incorporate additional metals such as nickel or chromium.

There are a number of processes for converting iron into steel. The basic oxygen process forces oxygen through molten iron, which helps to increase the internal temperature of the furnace and also combines with the carbon present, which is then released. Fluxing agents are added when the iron is molten, and a slag of impurities is formed.

Electric current is used to heat iron in the electric furnace process. This is similar to the method by which electric arc welding works (see **Connections**). The open-hearth process places the ingredients in direct contact with the fires that melt them. As with the basic oxygen process, oxygen is usually blown into the mix.

See also First Metalsmiths; Metals

FURTHER READING

Aitchison, Leslie. *A History of Metals.* New York: Interscience Publishers, 1960.
Ethical Metalsmiths, http://www.ethicalmetalsmiths.org
Fodor, R. V. *Gold, Copper, Iron: How Metals Are Formed, Found, and Used.* Hillside, NJ: Enslow Publishers, 1989.
Haedeke, Hanns-Ulrich. *Metalwork.* Vivienne Menkes, trans. New York: Universe Books, 1970.
Illinois Sustainable Technology Center, University of Illinois at Urbana-Champaign, http://www.istc.illinois.edu/info/library_docs/manuals/primmetals/intro1.htm
The Jefferson Lab Periodic Table of Elements, http://education.jlab.org/itselemental/index.html
ResponsibleGold.org, http://www.responsiblegold.org
United States Geologic Survey, http://www.usgs.gov

Modernism
(ca. 1890–1960)

Modernism was a style in the decorative arts that rejected tradition and ornamentation in favor of simplified lines, geometrically inspired forms, and minimal adornment. There have been many variations within modernism including stark geometries, simplified organic structures, and some very pared down references to neoclassicism, although the style was generally antihistoricizing. The resultant objects were meant to be both representative of their time and well-suited to their intended function.

Modernism originated in Europe, and its roots can be found in the first decorative styles that made a conscious break with the past. The initial stirrings of this change were found in some of the simplified forms of the **Arts and Crafts Movement,** although the reduced embellishment on such examples may have been the result of accommodating less well-trained craftsmen. Pared down forms were also evident in the work of some design reformers such as Christopher Dresser. The self-conscious antihistoricism was also a factor in many of the iterations of **art nouveau,** including the work of the Wiener WerkStätte in Vienna and the secessionist movement in Germany. The development of the design program of the **Bauhaus** mimics the Modernist movement in its evolution from hand craftsmanship to the prototype production of rational forms meant for machine production.

Modernism began in earnest at the 1925 Paris Exposition Internationale des Arts Decoratifs et Industries Modernes, in which only the newest designs were permitted (see **Art Deco**). Known as *style moderne* in France, this new aesthetic initially began in the luxury trades and included rich materials and sparse embellishment. Semiprecious stones, exotic woods, and time-consuming lacquers were used to adorn stark geometric forms in silver. French designers such as Jean Pufoucait were highly influential in the initial period of the style.

Early French modernism was influenced by cubism, and tended to include sharp angles and severe geometries. Scandinavian modernism was a warmer and more humanized alternative, with sparse but naturalistic ornament. This was welcomed more readily in countries that were slow to embrace the new styles, such as the United States. Companies such as **Georg Jensen** made inroads in markets abroad that were loath to accept the more extreme forms of modernism.

Modernism initially arrived in America in the late 1920s but did not gain much acceptance until the 1930s. Although the United States did not participate in the exposition of 1925, many Americans traveled there or read about it. American department stores took the lead in exhibiting objects in the new style just as Parisian stores had done in France. This was followed by museum exhibitions organized by members of the art and design fields in order to educate the American consumer. Retailers relied on advertising to create demand.

Early modernism in America was heavily influenced by European examples. Indeed, European emigrants or Americans who had studied in Europe designed much of the early American Modernist metalwork. The distinctly American versions of this style occurred later with the incorporation of skyscraper and streamlining motifs as emblems of progress and the machine age. Skyscrapers represented American advancement in the urban environment, and the use of skyscraper imagery in the decorative arts reached a high point in the late 1920s. Streamlining, a uniquely American version of the progressive aesthetic, incorporated rounded edges, silhouettes resembling aerodynamic trains, airplanes, or boats, and in some cases directional lines traveling across the surface implying movement.

The Great Depression of 1929 led to a slowdown in the metal trades and a general conservatism in the decorative arts. There was a significant return to familiar revival styles. In the modernism of the early 1930s elements of neoclassicism appeared as details on the simple forms, which lent itself well to inclusion in such a manner.

As the economy rebounded, new metal wares were inspired by symbols of progress, such as new technologies and transportation, and were meant to inspire confidence and optimism in consumers. Functionalism, designed to austerely reflect the purpose of an object, became a strong influence by the mid 1930s. Space-saving coffee sets that stacked or sat compactly on a tray were intended to appeal to urban dwellers with small apartment spaces.

World War II again affected the metal trades as factories were retooled for the war effort and raw materials were scarce. After the war, a kind of biomorphic abstraction, influenced by surrealism, was prevalent in metal design. This brought in a variety of asymmetrical amoeboid shapes. In the late 1950s, references to space travel also influenced design.

MATERIALS

At the end of the nineteenth century, new materials began appearing in **flatware** and hollowware. In some cases, they were used in the place of more expensive, traditional metals such as silver. In other cases, the new materials were used to provide a more durable and casual product. As a step down from silver, many manufacturers offered silver-plated metals as a way to have the appearance of silver without the cost (see **Gilding and Plating**). Companies were freer to try new designs in silver plate because it did not represent as significant of a material investment. However, silver plate still bore the more traditional look of silver and required the same hand polishing that solid silver did.

Alternatives to silver included chromium-plated metals, aluminum and, later, stainless steel. The changing role of women also aided the shift of metals within the home. Proposed to be a domestic technician of sorts, the early twentieth-century woman of the house was expected to be concerned with efficiency, health, and cleanliness, all of which were associated with the newer aluminum. Advertisers treated the role of housewife as a profession requiring scientific approaches and apparatuses. Aluminum cookware with an emphasis on functionality was marketed as the necessary equipment for such a role.

In the 1920s, aluminum began copying the forms of silver, trying to be beautiful as well as functional. In the decade following the Depression, aluminum was increasingly used out of the kitchen for entertaining as an inexpensive alternative to silver. In 1931 designer Russel Wright developed a line of spun aluminum designs that were very popular for less casual home entertaining (see **Lathe Work**). They were based on spheres and cylinders with an unadorned, modern appearance accented with wood or rattan. A few metalsmiths in Grove City, Pennsylvania, even chose to execute their one-of-a-kind designs in hand-hammered aluminum.

After World War II, aluminum was largely replaced by stainless steel for flatware, tablewares, and cookware. These new metals fit with the increasingly casual postwar lifestyle in which few had domestic help. The emergence of the dishwashing machine in 1949, in which aluminum and stainless steel could be safely washed, increased the desirability of the new materials as time savers.

Initially color was brought into silver wares with the semiprecious stones and exotic woods of French design. In other markets early plastics such as Bakelite and Catalin were used to add a touch of color from the late 1920s on. Some work out of Scandinavia utilized **enamels** and wood accents such as teak and walnut, which were copied by designers elsewhere. American enamellist Earl Pardon consulted at the Towle company in the 1950s. Other producers simulated enamels with durable paints.

DESIGNERS AND MAKERS

Many of the designers for the metalworking industry were trained as silversmiths abroad. Erik Magnussen was one such silversmith brought over from Denmark by the silver manufacturer Gorham. His designs were exceedingly popular in the late 1920s. They were characterized by simple structures with radiating, repeating lines that followed the curves of the overall form. One of his most well-known, and at the time least well-received, pieces was a cubist-style coffee set exhibited in 1927 called the "Lights and Shadows of Manhattan," which was covered with facets demarcated in silver, gilding, and **patination.**

German silversmith Peter Mueller-Munk also designed a number of pieces for serial production in the late 1920s and early 1930s. An exponent of handmade silver, Mueller-Munk came to America in 1926 and worked for **Tiffany & Company** as well as other firms before opening his own shop. It was his designs for industry that allowed him to survive the lean years of the Depression. One of his most well-known pieces is a chromium-plated pitcher for the Revere Brass and Copper Company known as Normandie, which resembles the funnels of a luxury cruise ship by the same name.

A major change in the metalworking industries was the rise of the professional designer in the 1920s. Such individuals might be called on to design anything from a building to furniture or silver flatware. Although many were trained as architects, they did not necessarily specialize in any particular medium, object type, or even scale. They were hired in large numbers by manufacturers to make their products appear more desirable to consumers. The influx of Europeans caused by both World Wars provided a large number of European-trained designers who were familiar with the new style of modernism. Some designers who made an impact on metalworking included the Hungarian-born Ilonka Karasz, Germans Elsa Tennhardt and Kem Weber, Finnish Eliel Saarinen, Austrian Paul Frankl, and American Donald Deskey. A great many more designers were anonymous. Major manufacturers of serial production silverware at the time included Reed & Barton, Towle, and International Silver Company.

Modernism greatly influenced the designs of the early postwar **Studio Movement** metalsmiths including Margret Craver. This was particularly true of

Scandinavian modern metalwork. A number of the early figures in the Studio Movement, including Ron Hayes Pearson and John Prip, worked in the smooth, organic style that was synonymous with Scandinavian metalwork at the time. Those men also had relationships with industrial concerns, making prototypes and designs for production. On a larger scale, modernism could be seen in the architectural work of German émigré Oscar Bruno Bach, who executed large-scale architectural commissions for wealthy American clientele that included machine-age figures in stainless steel and bronze.

The Depression and later wartime restrictions hurt the silver industry as well as individual makers. Craftsmen survived by augmenting their profession through teaching, consulting for industry, or working odd jobs.

Although no longer modern, aspects of modernism surface in some metalwork to this day. Certainly examples of it may still be found when purchasing tablewares and flatware.

See also Nineteenth Century

FURTHER READING

Braznell, W. Scott. "The Advent of Modern American Silver." *The Magazine Antiques 125* (1984): 236–41.

Bröhan-Museum, Dedo von Kerssenbrock-Krosigk, and Claudia Kanowski. *Modern Art of Metalwork: Bröhan-Museum, State Museum of Art Noveau, Art Deco and Functionalism (1889–1939), Berlin.* Berlin: The Museum, 2001.

The Dallas Museum of Art, Dallas, Texas, http://dallasmuseumofart.org

Heskett, John. *Industrial Design.* London: Thames and Hudson, 1980.

Kardon, Janet, and Rosemarie Haag Bletter. *Craft in the Machine Age, 1920–1945. The History of Twentieth-Century American Craft.* New York: H.N. Abrams in association with the American Craft Museum, 1995.

Modernism Magazine, http://www.modernismmagazine.com/

Stern, Jewel, Kevin W. Tucker, and Charles L. Venable. *Modernism in American Silver: 20th-Century Design.* Dallas, TX: Dallas Museum of Art, 2005.

Venable, Charles L. *Silver in America, 1840–1940: A Century of Splendor.* Dallas, TX: Dallas Museum of Art, 1995.

Nickel. ***See*** Metals

Niello

Niello is a blue-black or dark grey metal-based alloy that may be fused to the surface of other metals in order to create a visual contrast. "Niello" is the

European term for this technique, and is Italian in origin, derived from the Latin word for "blackened." Although recipes vary, the common ingredients are copper, silver, lead, and sulfur. These are melted together, then allowed to cool and solidify. The alloy is then pulverized and packed onto the surface of a prepared object. This object must be made of a metal with a higher melting point than the niello, such as copper, silver, gold, brass, bronze, or iron. It also usually has recessed areas on the surface into which the niello is packed. These recesses may be caused in any number of ways, including **casting, engraving,** and **etching.** Heat is applied uniformly to the object until the niello becomes molten and fuses to the parent metal. The excess niello is then ground down with fine abrasives until it is even with the parent metal and of similar shine. In some niello work, pieces of metal are dropped into the molten niello, where they become embedded.

The use of niello has been quite widespread over the course of history. Evidence of it dates as far back as **ancient Egypt.** Niello may also be found in metal work by most major civilizations prior to the **Renaissance,** when it reached a pinnacle of achievement in Florence, Italy. In more recent times, niello has been widely practiced in Turkey and Russia, with little use elsewhere. It was reintroduced to American metalsmiths by Phillip Fike during the **Studio Movement.**

See also Inlay

FURTHER READING

La Niece, Susan. "Niello: An Historical and Technical Survey." *Antiquaries Journal* 63 (1983): 279–91.

Nineteenth Century

The nineteenth century was a period of rapid change in the decorative arts. Machine production was increasingly used to make inexpensive goods for the home, and design became further separated from manufacture. Proponents of the Aesthetic Movement sought to curb the rampant eclecticism and turned to Japanese metal wares as examples. As a result, copper and its alloys found more places within the artfully decorated homes of the growing middle class. Even silver came within reach of most consumers through electroplating (see **Gilding and Plating**).

The period is sometimes referred to as the Victorian age because it was Queen Victoria who reigned in England at the time. During her reign from 1837 to 1901,

England was a wealthy nation undergoing large-scale expansion and industrialization. England was first to encounter the shift to primarily mechanized construction methods that changed the way metal goods were made in America and the rest of Europe. The term Victorian is also sometimes used to describe the eclecticism that was pervasive in the mid-nineteenth century.

In the early nineteenth century, large silver firms began to absorb and replace smaller, traditional silversmith's shops. They were able to invest large sums of money into mechanized processes that saved time and labor. Mass production techniques such as **stamping and die forming,** spinning, and electroplating made it increasingly possible to put metal goods into the homes of most individuals (see **Lathe Work**). One-of-a-kind objects in metal were reserved for the very wealthy, and their producers were increasingly in the minority. The rise of metalworking factories led to the demise of the **guild** system, which freed the new manufacturers from the traditional restraints within the crafts. The labor force became more mobile, and growth of production was unrestrained.

The divisions of labor that the factory setup demanded meant that individual craftsmen who were capable of making an item from start to finish were a rare exception. There was an increased separation of production from design. Further detachment was created through professional retailers. While the eighteenth century silversmith's shop usually designed, made, and sold their wares, these tasks were more and more performed by separate entities in the nineteenth century.

Styles in the decorative arts began to change more rapidly as communication and manufacturing processes sped up. Because the mechanically produced elements for these things could be made interchangeably, the number of stylistic configurations available to the consumer was staggering. The size of the middle class grew, and more new products were being turned out cheaply to satisfy their demands.

During the nineteenth century the reuse and commingling of styles from the past led to eclecticism in manufactured design. Metal objects for the home made between 1820 and 1870 were particularly diverse in their references. Revival styles, naturalism, and exotica including Far and Near Eastern art were the predominant sources. At the height of the Victorian period it was not uncommon to see motifs from as many as three styles coexisting on a single article. Revivals of such styles as the **rococo,** Gothic, **Renaissance,** and Colonial were frequently more fanciful than the original iterations. These gave way at the end of the century to a more restrained and academic eclecticism.

Trade with the Far and Near East, which increased with improved transportation methods, stimulated interest in those regions. Contact was furthered through the conquests of colonialism. Westerners viewed these cultures as idealized outsiders and valued their pre-industrial manufactures, which they supposed were more natural and less tainted by commercialism. Goods from exotic lands such as **India, Japan,** and the Middle East were popularized through their display at international expositions (see **Islamic Metalwork**).

It was this confusion of styles that the **Arts and Crafts Movement** and the Design Reformers protested. The Aesthetic Movement was also aligned against the new decorative tendencies, but without moral overtones regarding production methods. The Aesthetic Movement began in the mid-nineteenth century in England and had traveled to America by the time of the 1876 Philadelphia Centennial Exhibition. It championed art for its own sake and railed against the superfluous eclecticism of the day.

In an effort to bring more art into everyday life, the Aesthetic Movement's proponents strove to infuse household objects with artfulness. It was believed that carefully chosen objects placed within the home were able to reflect the owner's taste and worldliness while incorporating the modernity of mechanized manufacture. Aesthetic influences appeared in a variety of metalwork from cast iron stoves to silver tablewares. Styles popular within the movement included Japanesque, the pared down Gothic Revival, Islamic, and naturalism.

A nineteenth-century home's interior was supposed to provide an educational surrounding for the betterment of its inhabitants. Prior to the nineteenth century, metal goods in copper and its alloys were primarily utilitarian, not decorative. Beginning in eighteenth-century France, however, it became ever more acceptable for household furnishings to include or be made entirely from metals such as bronze and brass. Art brass was the name given to highly decorated brass goods for interior display. Brass became prevalent because of its inherent workability. Brass may be stamped, embossed, spun, used in thin sheets to cover other materials, and polished to a high golden-yellow finish.

As industrial manufacturers made metal objects in a variety of styles and at an affordable price, more and more decorative metal elements made their way into the middle-class homes. Entire pieces of furniture were soon being assembled from metal parts. Brass goods that might be found in a middle-class home included lighting devices, fireplace implements, accent tables and chairs, mirrors, clocks, and calling card receivers. Andirons became important decorative elements in late nineteenth century, as did the brass bed. Even cast iron found its way into Victorian garden furniture.

In 1853 Commodore Matthew Perry traveled to Japan with armed ships. This show of force resulted in Japan reopening their trade with the West the following year. Shiploads of Japanese decorative objects began to arrive in the West. It presented an alternative to historicism and was a major influence on the fine and decorative arts. The rage for Japanesque objects, Western articles made in a Japanese manner, began after art from Japan was first displayed at the London International Exposition of 1862. The color, texture, and asymmetrical natural imagery found on Japanese metal wares were highly influential.

The Japanese metal wares displayed at the 1876 Philadelphia Centennial Exhibition included mixed metal and **patinated** objects. Examples included both *mokumé gane* as well as the overlay of contrasting base metal alloys (see **Lamination Techniques**). The result was an increased interest in juxtaposed metals using copper alloys in the decorative arts. English metalsmiths were unable to copy the technique of incorporating copper or copper alloys with silver, as the assay laws did not permit it. America, however, had no such restrictions, and many silver companies began to produce Japanesque wares in silver that incorporated smaller elements of copper alloys. To further manipulate the color combinations, manufacturers also incorporated patinas. The Japanese wares in copper and copper alloys led to an increased acceptance of those materials for decorative purposes. The use of copper and its alloys also made it possible to make large decorative objects in metal that were not formidably expensive.

The nineteenth century also saw the rise of Britannia as a replacement for pewter. Unlike pewter, which has to be cast, Britannia was well-suited to mass manufacture techniques such as spinning or die forming. By the end of the nineteenth century, objects in copper, brass, bronze, and Britannia had found new places within the home.

During the nineteenth century, electroplated wares took the place of Sheffield Plate as less expensive alternatives to silver (see **Gilding and Plating**). In 1859 the Comstock Lode was discovered in Nevada. This and other new sources for the metal caused the price of silver to plummet. As a result, all but the poorest households had some silver-plated articles. Even the wealthy purchased electroplated wares for less important items. Inexpensive electroplating further fueled the invention of numerous specialty **flatware** implements.

Hand-crafted silver made on commission for the wealthy became the exception rather than the rule. Instead, standardized sets of goods like dinner and tea services were made in large quantities. Predetermined patterns replaced the silversmith's individual designs.

Articles made in silver began to be purchased in matching sets rather than acquired one at a time. Early in the century, tea sets generally included a teapot, sugar, and creamer. By mid-century it also included a coffee pot, hot water kettle on a warming stand, waste bowl, and matching tray. Other popular silver table articles included centerpieces, candlesticks, trays, display or serving plates, monteiths for cooling glasses, water pitchers, and casters for condiments. Small silver-plated trinkets were made is large numbers, including baby rattles, toilette sets, and inkwells.

In the second half of the nineteenth century, silver commemorative objects were made for most major occasions. Immense silver constructions were given in tribute to significant public events in unprecedented numbers.

See also Art Nouveau; Rococo and Neoclassical; Tiffany & Company

FURTHER READING

Bolger, Doreen. *In Pursuit of Beauty: Americans and the Aesthetic Movement.* New York: Metropolitan Museum of Art, 1986.

Brooklyn Museum. *The American Renaissance, 1876–1917.* Brooklyn, NY: Brooklyn Museum, 1979.

Burks, Jean M. *Birmingham Brass Candlesticks.* Charlottesville: University Press of Virginia, 1986.
D'Ambrosio, Anna Tobin, Nina Gray, Katherine A. Holbrow, and Gerri Strickler. *A Brass Menagerie: Metalwork of the Aesthetic Movement.* Utica, NY: Munson-Williams-Proctor Arts Institute, 2005.
The Victorian Society, the United Kingdom, http://www.victoriansociety.org.uk
The Victorian Society in America, Philadelphia, Pennsylvania, http://www.victoriansociety.org
The Victorian Web, http://www.victorianweb.org
Wright, S. M. *The Decorative Arts in the Victorian Period.* London: Society of Antiquaries of London, 1989.

Paktong. ***See*** Metals

Patina

A patina is the discoloration of a metal created through surface interaction with chemicals such as oxygen, chlorine, or acids. For example, rust is actually iron oxide, a patina that forms on iron when it interacts with oxygen. Patinas can be caused by time and natural exposure to the environment or induced through controlled exposure to certain substances. Sometimes a patina is valued for its aesthetic appeal or for its implications of antiquity. At other times, patinas are unwelcome for their corrosive effects. This is generally the case with silver sulfide, otherwise known as silver tarnish. The blue-green patina that is sometimes found on outdoor bronze statuary is a result of the airborne chemicals such as sulfur and water vapor corroding the bronze surface over time. In other instances, a patina may actually serve to protect a metal from further corrosion.

Copper and copper-based alloys form a wide range of patina colors particularly well. In less reactive metals like aluminum or titanium, the naturally formed metallic oxides give little color change and do not form patinas. Such metals must therefore be **anodized** to achieve different surface colors. The metals that we think of as being the noble metals (see **Metals**) tend to be corrosion resistant and thus do not form oxide-based patinas in their pure form. These include gold, silver, and platinum.

Simply heating a piece of metal will speed up its natural reaction with oxygen, thus creating a metallic oxide on the metal's surface. In the case of copper-based alloys, this may be used to achieve a wide spectrum of colors. In order to stabilize a patina achieved in this manner it is necessary to prevent further oxidation that would continue to alter the color. A paste wax or lacquer is usually applied to cut oxygen off from the metal's surface and thus halt any further oxidation.

Factors that may affect a patination process include: duration of exposure, temperature, humidity, nature of the chemicals involved, kind of metal, condition of the metal's surface, and application process. Common application processes for induced patinas include boiling immersion, wet application (brushing, spraying, or dipping), burying (in saturated dirt or woodchips), and fuming.

Metalworkers induce patinas for many reasons. It may be used to make a piece look older than it is by mimicking the natural patina that would otherwise develop with time. A dark patina may be used to emphasize a texture by patinating a piece of metal and then physically removing patina from the raised areas, which heightens the visual contrast between elevations. A patina may simply be used as a way of introducing a different surface color. Sometimes substances like paint are applied to mimic a patina, but unless there is a chemical reaction with the metal substrate these are not true patinas.

See also Armor and Arms; Enamel; Niello

FURTHER READING

Copper.org, "Finishes," http://www.copper.org/applications/architecture/finishes.html

Hughes, Richard. *The Colouring, Bronzing and Patination of Metals.* New York: Watson-Guptill Publications, 1991.

Young, Ronald D. *Contemporary Patination,* fourth edition. Escondido, CA: Sculpt-Nouveau, 1994.

Pearson, John
(active 1880–1910)

In many ways, John Pearson's life and work epitomizes the **Arts and Crafts Movement** in England. He is best known for his bold and imaginative **chasing and repoussé** work in copper. Pearson also participated in some of the organizations that have come to characterize the movement.

John Pearson most likely learned his skills as a metalworker and designer at one of the Home Arts and Industries Association classes. These were developed to preserve rural crafts at a time when industry was beginning to take over the production of all household goods. Crafts classes were believed to bestow a more morally uplifting experience than factory work provided. The Home Arts and Industries Association classes were also meant to provide extra income and a creative outlet for the working class that would keep them out of the pubs. By 1912 there were over 500 such clubs throughout Great

Britain. Metalworking as a subject of study was second in popularity only to woodcarving.

It is believed that Pearson worked for some time as a designer and decorator of tiles at William De Morgan's ceramic shop. De Morgan was a ceramicist and designer of stained glass windows as well as a founding member of the Arts and Crafts Exhibition Society, from which the movement got its name. The imagery used on De Morgan's tiles may have influenced Pearson's design aesthetic.

In 1888, Pearson was one of the five founding members of Charles Robert Ashbee's Guild of Handicraft at Toynbee Hall in London. Also the home of the School of Handicraft, Toynbee Hall had been developed as a form of experimental housing meant to give students in London greater contact with the poorer working class.

Pearson was the only member of the guild with training in metalwork and thus he designed nearly all of the early metal wares of the guild and trained the others in the metal shop to execute them. Pearson and the guild had 17 pieces included in the first exhibition of the Arts and Crafts Exhibition Society in 1888, all of which were metalwork and 15 of which were of Pearson's design. Pearson rarely worked on the designs of others, and in that way he embodies the arts and crafts ideal of joy in labor that was supposed to come from the freedom to execute every stage of an object's design and execution.

In 1890 John Pearson was expelled from the Guild of Handicraft for maintaining a second workshop outside the guild. This other workshop employed two men in addition to Pearson and supplied decorative metal wares to two of the most significant commercial outlets in the Arts and Crafts Movement: Liberty's and Morris and Co. Although Pearson was allowed to rejoin the guild shortly after being expelled, he resigned in 1892 to teach at the Newlyn Industrial Classes.

Newlyn was a fishing village in West Cornwall. The Newlyn Industrial Classes were started by designer John Drew Mackenzie to provide additional income for fishermen whose employment was dependent on the vagaries of the weather. Classes offered there included needlework, woodwork, and metalwork.

Pearson taught metalworking there throughout the rest of the 1890s. As copper was mined in the region, it was easily procured for use. Metal was the medium by which the decorative motifs of the Arts and Crafts Movement were most quickly spread in England. Copper, a very humble metal, was the ideal standard bearer of such a belief system. Its malleability and low cost added to the democratization of metalwork for amateurs.

John Pearson was best known for the bold relief that he achieved in his hammer-formed wares. The majority of work attributed to him is in copper, and it is believed that he never repeated a design. The motifs he used included ships, fish, peacocks, animals, and stylized natural forms such as plants and trees. Pearson created numerous chargers of boldly worked images but also made bowls, boxes, and other items of interior adornment. The hammer mark of the maker always remained as part of his finish, as a symbol of its handmade nature. The popularity of this kind of work is demonstrated in the fact that factories were mass-producing imitations of Pearson's style by 1900.

See also Nineteenth Century; Studio Movement

FURTHER READING

Brennan, Shawn. *Reflections: Arts and Crafts Metalwork in England and the United States.* New York: Kurland-Zabar, 1990.

Head Hand and Heart. "The Arts and Crafts Movement in Great Britain 1850–1915," http://www.artscrafts.org.uk

Karlin, Elyse Zorn. *Jewelry and Metalwork in the Arts and Crafts Tradition.* Atglen, PA: Schiffer Publishing, 1993.

Poikilia Decorative Arts, http://www.poikilia.co.uk/history.php

Persia
(ca. 3000 B.C.–640 A.D.)

Persia was the name of the region that is currently occupied by the countries of Iran and Afghanistan. This entry will focus on the people who lived there prior to the arrival of the Muslims in the mid-seventh century A.D. The word Persia comes from the Greek word for that region, Persis. The main inhabitants of the territory referred to themselves as Aryans. It is from the word Aryan that the country name Iran is derived.

The history of pre-**Islamic** Iran includes a series of conquering factions. With each successive political power, however, there developed a consistency within the arts that can be seen throughout the development of the metalwork from that region.

The original belief system of the Aryan tribes was polytheistic and nature oriented. Somewhere between the sixteenth and sixth centuries B.C., a prophet named Zoroaster preached a reformed, monotheistic religion that became known as Zoroastrianism. The principal deity is Ahura Mazda, which means Wise Lord. This was the religion of Persia prior to Islam. Zoroastrianism also incorporates elemental aspects of nature in its observances. The focus on the natural world that was a part of Zoroastrianism and the belief system that preceeded it is reflected in the metalwork from Persia in that the finest examples draw their inspiration from the natural world. Throughout the history of the region one finds depictions of animals that are both realistic and fanciful.

The land of Persia was abundant in natural resources, allowing for early working of a variety of metals. It is likely that the first metal worked was native copper. The first metalworking in the area may have been done as early as the seventh millennium B.C. Lost wax **casting,** smelting, and alloying were practiced around the middle of the fourth millennium B.C., and sophisticated metal forming was soon to follow. The use of silver and gold in significant quantities occurred late in the fourth millennium B.C. Ironworking appeared in the early second millennium B.C. Additionally, objects executed in copper alloys and lead have been found from ancient Persia.

The earliest vessels mimic previously developed ceramic versions. As metalworking grew in estimation, it began to influence ceramic forms in turn. In general, decorations were stylized and appear to have focused on decorative effect rather than strict representation.

The first unified group of people to occupy the region was the Elamites. Evidence suggests that they were active in the area as early as 3000 B.C. They made tools of arsenical bronze followed by tin bronze (true bronze). The production of bronze peaked around the second millennium B.C. and was slowly replaced by iron. Early iron implements were modeled on their bronze predecessors, despite the significantly different working processes (bronze was cast, iron forged). These implements included blades for axes and daggers. By the early first millennium B.C., only iron was being used for weapons, and bronze production focused on more decorative and symbolic items.

From this period there have been found cast bronze life-sized statues of deities and the royal family. Elamites also made bronze wall plaques, votive statues, and vessels for temples. Decorative imagery included plants, animals, and a figure known as the master of animals. The master of animals theme generally involves a human figure taming one or more wild animals. It is believed that this representation was meant as a metaphor for man's domination over nature and was common to the ancient art of southwest Asia.

In addition to casting, artisans of this period had mastered **raising,** soldering (see **Connections**), **chasing and repoussé,** and **engraving.** Some of the most technically advanced hollowware of the period includes raised vessels with animal decorations in chased relief on the sides. The animal bodies are in shallow relief but heads have been executed in high repoussé, creating a volumetric, rounded relief. Characteristics of this early metalwork persist throughout subsequent work from this region.

In the first millennium B.C. the Medes and then the Aryans began to move into the region. The Median civilization coalesced in the seventh century B.C. and was then conquered by the Aryans (Persians) in the middle of the sixth century B.C. Cyrus the Great led the Persians. He called his conquered region the Achaemenid Empire, after his ancestor, Achaemenes. This new group absorbed the indigenous style of metalworking, essentially that of the Elamites. It was further influenced by the work of the additional peoples they conquered as the empire grew. These collective influences became a cohesive Persian style. Their empire eventually stretched from **India** to Eastern Europe and down to North **Africa.**

The central focus of metalwork from the Achaemenid Empire was to glorify the monarchy. The court had objects made as gifts to foreign dignitaries, for use at banquets and as a way of preserving their wealth. The royal patronage led to the making of exceptional metalwork in gold and silver. Work developed both technically and stylistically.

Crafted objects included drinking vessels, bowls, dishes, and small gold and silver statues. Many of the silver implements created for the aristocracy were mercury **gilded** as well. Simpler versions were made of silver without gilding or of bronze. Inscriptions in cuneiform have been found on some vessels.

RHYTA

A rhyton is a straight or curved horn-shaped vessel that terminates in a secondary opening on its smallest end. The end is finished to resemble the head or torso of some kind of beast or person. Rhyta were made in the region of the Mediterranean and Middle East at least as far back as the first millennium B.C. They were used at Persian court from the Medes to the Sasanians.

Rhyta have been found in ceramic, silver, and gold. Earlier examples have the creature's head projecting straight out from the main axis of the vessel. Later rhyta curve at the end, putting the head at a right angle to the length of the vessel. The use of Rhyta was widespread, and the animals depicted were various, including cats, rams, horses, camels, bulls, and mythical creatures. These might have had different connotations depending on the culture using the vessel.

Alexander the Great conquered the Achaemenid Empire in 331 B.C. Large quantities of gold and silver were taken from the Persian treasuries and sent back to Greece until Alexander died in 323 B.C. When the vast empire Alexander had conquered began to break down, one of his generals, Seleucus, assumed control of Persia. The reign of his successors was known as the Seleucid Empire and lasted until 250 B.C. In metalwork this period was marked by the imposition of Hellenistic influences on the extant Persian styles. Silver appears to have been particularly popular during this period.

Instability in the region led to a takeover by a nomadic group called the Parthians. Parthian rule lasted from around 250 B.C. to 224 A.D. Their lands included a large portion of the Silk Route. The Silk Route, or Silk Road, was a series of passages connecting the Far East with Europe. Goods and ideas moved great distances on these roads, and the Parthians gained income by taxing the travelers. The products and belief systems that passed through also influenced them.

The Parthians lacked a strong centralized government. The diverse local customs make it hard to discern a cohesive style of metalwork from this time. The surviving objects from this period reflect eclectic influences, including Persian, Greco-Roman, and Far Eastern cultures. Dishes that were exuberantly covered with evenly distributed, centralized, and stylized elements epitomize these traits. Architectural motifs show up on some pieces. Imagery associated with the Greek god Dionysus, such as wild cats and grape vines, also appear on a number of drinking vessels. The popularity of his cult had spread with the Hellenizing influences that followed the conquest of Alexander the Great.

After half a millennium under other rulers, Persia was returned to the Persians with the rise of Ardashir. He conquered the Parthians in 224 A.D. and began the Sasanian dynasty, named for his grandfather Sasan. This was a golden age for Persian metalwork, which peaked around 550 A.D.

The Sasanians were contemporaneous with the **Byzantine Empire,** which existed to the west of them. They also had interactions with **China** to the east, and continued to benefit from the trade route that passed through Persia. Although they outwardly rejected Roman influences, some Hellenistic style hade become part of the region's visual vocabulary.

The most accomplished objects of the Persian metalsmithing production were silver vessels. As with the Achaemenids, these objects were used at banquet, given as gifts, and hoarded as wealth. Early bowls resembled simplified Parthian vessels with centrally located chased and repoussėd busts or hunting imagery featuring the bowl's owner.

It is believed that during the fourth century, silver production came under tight control of the monarchy. All the crafts were unified is a system similar to medieval **guilds.** This led to more uniformly produced work with the king represented in most figural ornaments. The king was generally shown hunting, banqueting, or sitting on the throne. Sometimes dancing females were depicted instead. Additional, recurring motifs included plant and animal forms. Animals could be real or fanciful, such as winged dragons. The mythical animals featured in Sasanian metalwork are thought to have a variety of meanings, from guardians to evil spirits. Aspects of Greek mythology and drama are also depicted. Some vessels were inscribed with their weight, possibly for tax records.

Precious metals were used to promote the strong centralized government through imagery and grandeur. Silver vessels were the most common, and sometimes they were gilded. Gold and bronze vessels occur less frequently. Such vessels took a variety of forms, including round or oval dishes, sometimes with pronounced lobes. Ewers and vases also survive in number in silver and bronze. These were made through both raising and casting. Decoration might include engraving, chasing and repoussé, **niello** inlay, or gilding. Gilding on such works is usually applied to the background of the image, rather than the raised imagery. Sometimes the chased elements were executed separately and soldered on. Bronze examples tend to exhibit less ornamentation. Bronze heads created to resemble various rulers have also been found from this period.

At the peak of Sasanian metalworking, repoussé created highly developed relief surfaces on vessels as well as objects in the round. A number of hollow-formed animal statuettes with chased details have survived. The Sasanians also utilized **enamels** and produced some fine examples of cloisonné. Much of their enamel work copied decorative forms found in other media, such as textiles.

By the middle of the seventh century the Sasanian Empire had begun to fall apart from within. The arrival of Islam around 640 marked the end of the Persian Empire. However, styles and motifs of the Sasanians were influential and largely continued in the early years of Islamic metalwork from that region.

See also First Metalsmiths; Middle Ages

FURTHER READING

Allan, James W., and Brian J. J. Gilmour. *Persian Steel: The Tanavoli Collection.* Oxford: Oxford University Press for the Board of the Faculty of Oriental Studies, University of Oxford and the British Institute of Persian Studies, 2000.

Ferrier, R. W. *The Arts of Persia.* New Haven, CT: Yale University Press, 1989.

Freer Gallery of Art and Arthur M. Sackler Gallery, Smithsonian Institution, Washington, DC, http://www.asia.si.edu

Ghirshman, Roman. *The Arts of Ancient Iran from Its Origins to the Time of Alexander the Great. The Arts of Mankind.* New York: Golden Press, 1964.

Harper, Prudence Oliver. *Silver Vessels of the Sasanian Period.* New York: Metropolitan Museum of Art, Princeton University Press, 1981.

Ross, E. Denison. *Persian Art.* London: Luzac and Company, 1930.

The State Hermitage Museum, St. Petersburg, Russia, http://www.hermitagemuseum.org

Yarshater, Ehsan, ed. *The Cambridge History of Iran Volume 3: The Seleucid, Parthian and Sasanian Periods.* 2 vols. Cambridge: Cambridge University Press, 1983.

Pewter. ***See*** Metals

Platinum. ***See*** Metals

Pre-Columbian
(ca. 2000 B.C.–1540 A.D.)

The ancient American civilizations that flourished prior to European arrival are generally referred to as pre-Columbian. Successive tribes of native Indians, many of whom made remarkable metalwork, once occupied Latin America in the areas that are now Mexico, Central America, and South America. Regions are being referred to by their present country names for ease of use; these delineations did not exist prior to the European invasion.

Christopher Columbus arrived in the New World in 1492. Following his lead, a number of other explorers came from Spain seeking riches. By 1600 the advanced civilizations that had existed from Mexico to the Andes in South America had been largely taken over. Gold and silver was melted down and cast into ingots to be shipped back to Spain, where it was tuned into currency. A little information has been passed down as to what the Spanish saw, including accounts of gold-covered furniture, golden gardens with corn, flowers, and animals all executed in the glowing metal, and whole rooms and exteriors of buildings sheathed in gold foil.

At the time of the European conquest, the two major civilizations in the area were the Inca and Aztec. Gold was commonly given as tribute to the rulers of both groups. In both societies, the central leadership controlled the gold, which was largely located at their respective capitals. The central location made it easier for the Europeans to strip each culture of its metals.

Metalworking may have started as early as 2000 B.C. in the Andes Mountains of South America. From the area that is now Peru the technology of metalworking gradually spread northward toward Central America. The pre-Columbian people primarily worked in copper, gold, silver, and their alloys. Additional metals such as tin and platinum were used with less frequency. All the metals they made use of were nonferrous, that is, without iron.

The initial development of forms in metal was considerably different than that of early European metalworking. The ancient Latin American interest in precious metals was not involved with monetary worth, and these metals were rarely used in trade. Rather, the metals had spiritual and religious significance. Gold was seen as an element aligned with the sun god and, by extension, the ruler who they believed was descended from the sun. Precious metals were used in religious rituals as symbolic offerings and were buried with the nobility.

Only the nobility, chiefs and priests, were permitted to wear gold ornaments. However, their interest in gold was symbolic, as they believed that gold emanated the energy of the sun. The function of gold was not as a commodity but rather as a bearer of meaning. The use of gold communicated the power structures and values of the culture. The Inca believed that the first member of their ruling class had been the son of the sun. Gold was affiliated with the male deity and silver with the female. Gold and silver also conveyed personal status in life and death.

Because of the destruction of the metalwork of the pre-Columbian cultures, what we know about them today comes largely from a few burial sites. Metalworkers generally had a high social status in pre-Columbian civilizations. Some goldsmiths were only permitted to work in gold and other labor was forbidden to them. As gold was normally under the control of those who governed, so were the metalworkers. The trade was generally masculine and familial, passing from father to son.

The first metal used by the Latin American Indians was most likely gold. This use appears to have begun around 1500 B.C. in Peru. It was available in alluvial deposits in rivers in much of the region. The first things the natives could have done with the gold they found in rivers would be to pound on them and grind them with rocks. The next stage of metalworking involved heat applications, which included **annealing** for greater malleability. This led to an increased ability to shape the metal through force. Some kind of incising may have taken place as well, similar to **engraving.** Greater heat application would have led to molten metal, meaning the ability to solder (see **Connections**) and **cast.**

Casting was a widespread technology, from Peru to Mexico, by the first millennium A.D. Development probably progressed from simple open molds to two-piece clay molds. This was how they made small decorative items as well as blades for axes and knives. The lost wax casting method developed later, and several groups made elaborate, false **filigree** by forming objects of wax wires and then casting them. With the ability to melt metals also came alloys, in which pre-Columbian metalsmiths added copper to gold or silver to increase their rigidity for hammer forming and lower their melting temperatures for casting. Silver was also alloyed with gold in order to produce solder. *Tumbaga* was the name the Spaniards gave the copper and gold alloy used by the native peoples. Whether they had been hammer formed or cast, these gold-based alloys were given a skin of relatively pure gold by depletion gilding (see **Gilding and Plating**). Using heat and the application of acidic plants and local minerals, the metalsmiths could remove the copper oxides and silver oxides from the surface of gold alloys.

What the metal workers of these cultures could achieve with relatively limited tooling was quite amazing. They had no iron but used stones for hammers, chisels, and polishing. They melted metals in copper crucibles and hammered sheet metal over stone and wood matrices. Methods of ornamentation included **chasing and repoussé,** engraving, and the **inlay** of metals and stones. In addition to soldering they also used tab and staple connections. Larger pieces were made by sheathing wood or bone substructures in gold foil.

Some bronze was used by South American cultures, but it does not appear to have been as significant or entrenched in their society as copper, gold, and silver. In general, the more northerly regions of Mexico, Central America, and Colombia seem to have cast most of their metal objects. In South America, particularly in the Peruvian Andes, the forming of sheet metal was more common.

The objects made in metal included vessels for eating and drinking, sun discs, votive figures and animals, items for personal adornment, rattles, furniture finials, and death masks. Strictly utilitarian items like tools appear to have been made less frequently. Creatures that show up repeatedly in metalwork from these cultures include jaguars, frogs, alligators, snakes, and birds. In general, forms were broad with bold designs and little detail.

SOUTH AMERICA

The greatest metalworking activity was in the Andes Mountain region. Peruvian metalsmiths tended to work gold into sheet and then produce objects by shaping the sheet, using casting only for details that were soldered on in later stages of production. They hammer formed sheet metal over precarved wood and stone forms, then joined the elements with solder, creating complex and hollow sculptural figures. These were both human and animal in reference. The Moche, Chimú, and Inca people all produced objects this way. It is thought that the Moche (ca. 200 B.C.–700 A.D.) may have been the first people in the Americas to develop lost wax casting, for which they used copper alloys. The tombs of Moche chieftains have been found filled with gold and silver.

Craftsmen in the region of Ecuador developed a method of sintering platinum, both with itself and with gold. In sintering, small particles of a metal can be heated and joined together with pressure at a temperature lower than its melting point. This enabled the working of platinum, which has a very high melting temperature, prior to the technology of achieving that temperature. This technique was lost during the Spanish incursion. Platinum was not used in Europe until the nineteenth century.

The Tiahuanaco (ca. 200 B.C.–600 A.D.) occupied parts of Peru and Bolivia. Some of the items ascribed to them include **raised** beakers with masklike relief repoussé on the sides. Beakers are tall, handleless drinking vessels with either straight or slightly flaring walls. They also made work that exhibits wrigglework engraving. The Chimú (ca. 1350–1465) similarly made raised beakers in large quantities, some of which look like figural human heads. The Chimú also soldered and cast ceremonial knives from precious metals, some of which alternated gold and silver from the process of casting in place. In this procedure, a wax model of an object is made and then cut into two or more pieces. One part is cast in the metal with the higher melting temperature, cooled and cleaned. Then the adjoining piece of wax is affixed to it and the lost wax process is repeated; the wax is melted out and a metal with a lower melting temperature replaces it and diffusion-bonds to the adjoining, precast metal element. This bimetal casting was also done in Mexico.

By the time the Europeans arrived, the Incas (ca. 1150–1532) had been extending their empire for a hundred years. They controlled land that extended thousands of miles along the Peruvian coast and included millions of people. Their metalworking was also fairly advanced. They had built on what previous societies had done and added the technique of metal inlay to their repertoire. They continued the traditions of ceremonial vessels, personal adornment, and hollow-formed, freestanding figures. Some Europeans reported seeing such figures made in life size along with replicas of plants and animals.

CENTRAL AMERICA

In general, metalworking techniques seem to have moved from Colombia up through Central America to Mexico. Most metalwork there was made from *tumbaga* through the lost wax casting process. A number of places without readily available gold deposits of their own turned to Panama for their supply, including Costa Rica and the Yucatan Peninsula.

MEXICO

The working of gold came later to Mexico, around 1000 A.D. The work in metal became sophisticated quickly, however, and may have been aided by traveling artisans. A number of discs with elaborate, narrative repoussé have been found in Mexico, but most of the pieces were cast items of personal adornment.

The Aztec (ca. 1150–1521) emperor Montezuma II attempted to forestall the arrival of Spaniard Hernán Cortés by sending gifts to him of gold, silver, and stonework. Cortés sent these back to Spain, where they were wondered at by the court and then and apparently melted down. Cortés was enticed by the riches rather than assuaged and went on to loot the Aztecs and ship their gold and silver back to Spain as ingots.

See also Africa; First Metalsmiths

FURTHER READING

Dumbarton Oaks, Research Library and Collections, Washington, DC, http://www.doaks.org

Emmerich, André. *Sweat of the Sun and Tears of the Moon: Gold and Silver in Pre-Columbian Art.* Seattle: University of Washington Press, 1965.

Jones, Julie, ed. *The Art of Precolumbian Gold: The Jan Mitchell Collection.* Exhibition catalogue. New York: Metropolitan Museum of Art, 1985.

Lothrop, S. K. *Essays in Pre-Columbian Art and Archaeology.* Cambridge, MA: Harvard University Press, 1961.

Museum of Gold, Bogotá Colombia, http://www.banrep.gov.co/museo/eng/home4.htm

National Museum of Anthropology, Mexico City, Mexico, http://www.mna.inah.gob.mx

Peabody Museum of Archaeology and Ethnology, Harvard University, Cambridge, Massachusetts, http://www.peabody.harvard.edu

Wardwell, Allen. *The Gold of Ancient America.* Greenwich, CT: Distributed by New York Graphic Society, 1968.

Raising

Raising is a technique by which flat pieces of metal are hammered into volumetric shapes through compression. Raised vessels have been found in ancient Sumerian graves in the Middle East dating from around 2500 B.C. It was the primary method by which vessel forms were created from sheet metal prior to the Industrial Revolution. A number of different raising and hammer-forming techniques exist but have not changed in thousands of years.

What is typically thought of as raising today is specifically known as angle raising. In angle raising, the piece of sheet metal is usually cut into a circle and **annealed.** The disc is placed with its center resting at an angle upon a stake. A stake is a piece of hard material, usually steel, which comes in a variety of shapes and is used to support metal while it is being worked. The metal being raised is struck just beyond the point where it rests on the stake, thus causing the metal to compress. This is done in a pattern of concentric circles, slowly spiraling out from the center of the piece to its edge. After each round of raising, the outer circumference of the disc has been reduced and the walls of the vessel have become more vertical. The sheet metal must be annealed after each round. Raising continues until the walls of the vessel have achieved the desired shape.

Dutch raising is similar to angle raising but is done in steps from the outer edge, inward. In crimp raising, regularly spaced ruffles that gently radiate from the center are lightly hammered into the sheet metal. These are then compressed into the body of the vessel through hammering as in the process of angle raising. Stretch raising requires the use of a much thicker piece of sheet metal. The metal is struck with a bulbous hammer while it rests on a flat stake or anvil. This is generally done from the center outward in concentric circles, sparing the edge. The thinned metal develops a greater surface area, and as it is pinched between the hammer and stake, greater volume is formed. Unlike other kinds of raising, stretch raising preserves the diameter and thickness of the metal's edge.

To start the raising process, sometimes a piece of metal will be sunk first. Sinking involves hammering a piece of annealed sheet metal into a rigid depression or sandbag with a large, rounded hammer. This is a quick way to develop shallow forms, but the process thins the metal and is therefore not generally used when much depth is wanted.

After raising, the process of planishing is usually used to make the vessel's surface more regular. The finished form is supported by a stake that fits the contour being worked on (this may require several stakes for one piece). A hammer with a flat, round face and a polished surface is used to gently tap and compress the surface of the metal against the stake below. This is usually done over the entire piece several times and may result in a flawless surface if done correctly.

See also Chasing and Repoussé; Forging; Lathe Work; Stamping and Die Forming

FURTHER READING

Hurt, William, Age of Armour, (raising demonstration), http://www.ageofarmour.com/education/raising.html

McCreight, Tim. *The Complete Metalsmith: An Illustrated Handbook,* revised edition, Worcester, MA: Davis Publications, 1991.
Untracht, Oppi. *Metal Techniques for Craftsmen; a Basic Manual for Craftsmen on the Methods of Forming and Decorating Metals,* Garden City, NY: Doubleday, 1968.

Renaissance
(ca. 1400–1600)

The Renaissance represented a new worldview that began in Italy in the fourteenth century and spread slowly throughout the rest of Europe over the following two hundred years. The word renaissance means "rebirth," and this time period was seen as a revival of classical ideals. The Renaissance was a renewal of the secular, humanist-centered philosophies of classicism. This was in contrast to the more spiritually based **Middle Ages**, which had preceded it. Artists and philosophers of the age revisited the work of **ancient Greece** and **ancient Rome** for inspiration. Archaeological excavations in Italy helped fuel the interest in classicism. Renaissance artists strove toward the representation of physical reality rather than religious symbolism. It was a period of great inquiry and learning.

In the early Renaissance, towns that had grown into self-governing entities at the end of the Middle Ages were growing in importance and prestige through trade and manufacture. A greater variety of objects were available to a growing middles class as well as the aristocracy. Increased demand was met through the highly organized **guilds.**

One of the major revolutions of the period was the invention of a moveable type printing press. Around 1440, Johannes Gutenberg in Germany developed this method of printmaking. Prior to this breakthrough, all books in Europe were either hand copied or printed from carved wooden blocks. The printing press allowed for a much more rapid spread of ideas. Design books full of representations of metalwork helped to disseminate Renaissance styles.

The decorative motifs used during the Renaissance were largely derived from the surviving art and architecture of ancient Greece and Rome. These were usually made of stone and ceramics, thus Renaissance metalworkers had to adapt the design elements. Decorative subjects included figures from Greek and Roman mythology, scrolls, foliage, *putti* (pudgy, winged boys), grotesques, interlace, and the quatrefoil. A quatrefoil is an abstract floral form with four equal-sized petal shapes radiating from an open center. Ornamentation was frequently stacked in horizontal bands. It was above all a style that valued human intellect. Being aware of the classical subjects used and their allegorical meanings meant that you were an educated person.

SILVER AND GOLD

In the later Renaissance, silver demand increased as the merchant classes expanded. The nobility continued to use silver and gold tablewares as a way to store and display their wealth, to exhibit power and piety, and to make political connections. Early Renaissance precious metalwork in Italy followed the fashion for classical references. This did not influence the rest of Europe as much as Mannerism did in the sixteenth century. Mannerism was a late Renaissance stylization. Its forms tended to be more complex and elongated than early Renaissance decorative motifs.

Regional variations occurred as well. Germanic pieces tended to hold onto Gothic references longer. Portuguese metalwork incorporated African themes, which referred to the source of their gold. Lavish accoutrements for church ceremonies dominated Spanish production in precious metals from this period, fueled by wealth from the New World (see **Pre-Columbian**).

Antwerp and Nuremburg were known for their goldsmiths' work at the time. Artisans in those cities produced many elaborate mounts for exotic specimens such as ostrich eggs, nautilus shells, and coconuts. These were generally turned into cups with lavish settings, not meant for use but display only. Gold vessels incorporating certain precious stones were believed to impart particular magical properties to the user. The medieval tradition of making reliquaries also continued.

Gold was sometimes used to produce diversions for the wealthy and powerful, such as automata. An automaton is a machine meant to perform certain movements through its own powers. Such objects of wonder were even more highly valued when executed in precious materials.

CABINETS OF WONDER

In such an age of scholarly study, people assembled collections of objects for examination and diversion. The physical evidence of human and natural history were explored and kept in personal treasuries. These were called *wunderkammer* (wonder cabinets) or *kunstkammer* (art cabinets) in German, *cabinets* in French, and *studiolos* in Italian. These were a kind of personal museum meant to reflect the wealth and intellectual accomplishment of the owner. A successful *wunderkammer* required discrimination and education to assemble and generally included examples of both the natural and man made. Particular emphasis was given to things that were either rare, exotic, or from classical antiquity. In metalwork this included gold and silver mounted cups of hard stones or exotic materials as well as cleverly fabricated automata or instrumentation.

COPPER ALLOYS

Copper was primarily used for domestic vessels, such as water jugs, and tended to be more popular in Italy than in northern regions. The copper alloys of bronze and brass were used extensively throughout Europe. Lost wax bronze **casting** was used in Renaissance Italian statues and sculptures ranging from the life-sized to intimate scale. Sculptors such as Donatello and **Benvenuto Cellini** were known for their work in this medium. Like Cellini, many bronze sculptors had been trained as goldsmiths. The cast bronze baptistery doors by Lorenzo Ghiberti from the first half of the fifteenth century continue to inspire awe.

Domestic cast bronze articles included mortars, inkwells, candelabra, and candlesticks embellished with the heavily modeled designs. Mortars were particularly valued utensils for grinding spices and were sometimes given as wedding presents. Other domestic metalwork might include pots, crucibles, smoothing-irons, basins, weights, chamber pots, wall hooks, hand lanterns, spigots, warming pans, and table rings for supporting dishes.

HERRENGRUND CUPS

Herrengrund was a historic mining town in Hungary that produced famous copper cups in the sixteenth and seventeenth centuries as the result of a seemingly magical transformation. The people of Herrengrund discovered that if they put iron in some of the town's water, it appeared to transform into copper! What was actually happening was a process known as cementation. The local ground water was naturally rich in copper, and the introduction of iron created a small electric charge that precipitated the copper out of solution to plate the iron. Cups were made from this copper as showpieces, modeled after elaborate silver and gold examples. They frequently were chased, engraved, or enameled with images or inscriptions referencing their unusual origins.

During the Renaissance, chandeliers slowly moved from churches into the homes of nobility and finally to the wealthy merchant class. They also became more elaborate over time. Candlesticks were produced in large numbers with regional design variations. During the Renaissance, the market for candlesticks grew considerably, and artfully designed ones became part of the luxury trade. Figural vertical elements, with each arm extended to support a candle, were popular in the Low Countries. The figures represented could be mythical, religious, or regional caricatures. The baluster candlestick form was popular across Europe. Utilitarian forms produced for display purposes in brass, bronze, and pewter grew as the middle class mimicked the silver-laden sideboards of the wealthy.

Extraordinary objects such as scientific instruments were manufactured at a higher level than domestic articles because they required greater skill and an understanding of the phenomena that their instrumentation was to measure. These included clocks and sundials. Brass and gilded bronze were used for showpieces, which were made by specialists.

Venetian brass and bronze wares, such as candlesticks and basins, were highly influenced through exchange with **Islamic** cultures. Extensive trade networks made Venice prosperous and exposed its inhabitants to a wide array of stylistic influences. In the sixteenth century, Islamic designs in brass wares from Venice appear, suggesting that some Islamic peoples had come there, possibly from **Persia.** These wares characteristically included complex metal **inlays.** This style was later adopted by local craftsmen who incorporated classical decorative motifs as well.

DINANT AND NUREMBURG

The town of Dinant, in modern-day Belgium, was a major producer of hammered brass articles from the twelfth through mid-fifteenth century. Because of the profusion of Dinant's output, such wares are sometimes referred to as dinanderie. Although basins were probably the most well-known items, domestic and religious lighting and pouring vessels were also made. After the attack of Phillip the Good in 1466, the Dinant metalworkers dispersed throughout Europe.

In the following century, Nuremburg, Germany, became a major producer of copper, brass, and bronze goods. Part of Nuremburg's rise to prominence may have been aided by the unusual organization of trades within the town. Rather than a guild system, the town council regulated the town's trades. Occupations were divided into free trades that anyone could participate in and the more stringently regulated oath bound trades, of which metalworking was one. The Nuremburg makers of basins, the so-called basin beaters, were particularly well known throughout Europe. Basins with repoussé decorations featuring religious, allegorical, or abstract motifs were generally used to decorate walls and sideboards. The weight makers of Nuremburg were also very important. Weights were made in the forms of nesting buckets, which were sold throughout Western Europe.

Nearby Augsburg was also known for its clockmakers and goldsmiths, who were equal to the task of producing royal gifts.

PEWTER

Pewter was used for domestic vessels but was also swept up in the Renaissance interest in conspicuous wealth. In the sixteenth century, highly ornamented display pewter became popular for the decoration of homes and sideboards. It displaced brass wares of this kind because cast pewter could achieve a higher degree of detail than repousséd sheet brass. Display pewter was first produced in Nuremburg, Germany and Lyon, France.

François Briot was a particularly famous die cutter for this kind of pewter. The detailed relief decorations that he executed were full of scholarly references. His most famous work, the Temperancia Dish, has been copied in pewter and ceramic and was also the electrotype model for the modern-day Women's Singles Trophy at Wimbledon (see **Electroforming**).

IRON

Iron work became influenced by Renaissance ideals later in the period. The most complex wrought iron work of the period was found in Germany, France, and England where it was used for screens, gates, balcony rails, and domestic hardware. The quatrefoil motif was particularly popular in iron. During the Renaissance, ironwork became less architectonic and was executed more in the manner of other metalwork, including such motifs as scrolls, foliage and interlace. German blacksmiths made particularly challenging pieces in the sixteenth century.

As it became increasingly affordable wrought iron was used for a wide variety of implements including candlesticks, fireplace and cooking utensils, and locks with keys. The wealthy commissioned elaborately carved and etched locks for doors and cupboards. Small chests of iron were made for decorative purposes or to transport small amounts of money or jewelry. A famous form of these miniature caskets was the Michel Man, named for a late Renaissance metalworker from Nuremburg.

Iron was also cast into busts and architectural elements and sometimes embellished by gilding or painting. Domestic uses for cast iron included cast iron mortars, pots, stove panels, grave markers, and firebacks that protected the back wall of a fireplace.

ENAMEL

In the fourteenth century the **enameling** center Limoges, France, collapsed due to political strife in the region. Following a resurgence in the late fifteenth century, the city resumed its dominance of enameling that had begun in the Middle Ages. The practitioners there excelled at the techniques of *basse taille* and cloisonné. During the Renaissance, a more painterly technique of applying the enamel also became popular. This followed the trends in fine art painting at the time. Designs were regularly taken from printed books, and classical mythology began to replace religious themes. Enamels were used on plaques, horse fittings, vessels, jewelry, and tomb covers.

Plique-à-jour and *en ronde bosse* enameling were both developed during the early years of the Renaissance. *En ronde bosse* translates into "in round relief." This refers to enameling executed on three-dimensional objects—those surfaces that have been chased and repousséd or cast in the round. Enamel was frequently used on the figures in reliquaries and jewelry to make them appear more lifelike.

In the sixteenth century, a type of enameling called *grisaille* became popular. Grisaille comes from the French word for grey, which is *gris.* In grisaille enameling the dominant colors are black, white, and grey. It is usually executed by building up layers of white enamel over a black background to achieve an image composed of grey tones. Subject matter ranged from religious to classical mythology and frequently incorporated a customer's coat of arms.

See also Byzantine Empire; Flatware

FURTHER READING

Bagatti Valsecchi Museum, Milan, Italy, http://www.museobagattivalsecchi.org
Belozerskaya, Marina. *Luxury Arts of the Renaissance.* Los Angeles: J. Paul Getty Museum, 2005.
Gruber, Alain Charles, and Margherita Azzi Visentini. *The History of Decorative Arts. Renaissance and Mannerism in Europe.* New York: Abbeville Press, 1994.
Keith, D. Graeme, and Charles Avery. *The Triumph of Humanism: A Visual Survey of the Decorative Arts of the Renaissance.* San Francisco: Fine Arts Museums of San Francisco, 1977.
The Society for Renaissance Studies, Great Britain, http://www.rensoc.org.uk
Syson, Luke, and Dora Thornton. *Objects of Virtue: Art in Renaissance Italy.* Los Angeles: J. Paul Getty Museum, 2001.
Thornton, Peter. *The Italian Renaissance Interior, 1400–1600.* New York: H. N. Abrams, 1991.
Victoria and Albert Museum, London, England, http://www.vam.ac.uk
Wilson, Carolyn C. *Renaissance Small Bronze Sculpture and Associated Decorative Arts at the National Gallery of Art.* Washington, DC: National Gallery of Art, 1983.

Revere, Paul
(1734–1818)

Paul Revere is best known for his efforts in the colonial resistance to England prior to the American Revolutionary War. It is frequently overlooked that he was also a goldsmith, merchant, and entrepreneur whose business interests were closely intertwined with the establishment of the new nation.

Paul Revere's father, Apollos Rivoire, was a French Huguenot who had immigrated to Massachusetts around 1715 (see **Paul de Lamerie**). Rivoire apprenticed as a goldsmith under John Coney (most goldsmiths worked primarily in silver). Around 1729, Apollos Revere opened his own shop, married Deborah Hitchbourn, and changed his name to Paul Revere. Their first son was given the same name and trained as a goldsmith under his father. When the elder Revere died in 1754, the goldsmith shop remained open in his widow's custody until young Paul was old enough to legally assume the responsibility. In 1756 Paul served in the military against the French. When he returned the following year, Paul took control of his father's shop where his younger brother, Thomas, aided him.

The **rococo** style was popular when the younger Paul began his apprenticeship. The shop produced work in that style with a certain amount of restraint in the ornamentation. Paul's work was well made and equal to that of any of his peers. In the years before the Revolutionary War, as he was proving himself as a goldsmith, Revere tended to take more commissions and work on specialty pieces. After the war he would restrict the output of his shop somewhat. As with most goldsmith shops he had a number of apprentices, journeymen, and hired employees over the years including some of his sons and other relatives (see **Colonial America**).

In the years preceding the Revolutionary War, economic circumstances were unstable. In order to weather such conditions, Paul diversified the services that he offered. Revere used his skills as an **engraver** to produce copper printing plates for publications, trade cards, and bookplates. He offered dentistry services including the cleaning of teeth and the making and installation of false teeth. At the time it was also common for goldsmiths to act as retailers for merchandise shipped from England.

Social and professional connections were very important to the trade of colonial silversmiths. Because of the inherent value of their materials they had to be trustworthy, and clients were acquired through such relationships. Young Paul had inherited his father's business acquaintances and was well connected within the metalworking community. He also joined a number of clubs and organizations including the St. Andrews Masonic Lodge, the Sons of Liberty, and the North End Caucus. By taking a part in the affairs of his community, Paul expanded his business networks.

One article made by Revere stands prominently as a lasting tribute to his political ideals: the Sons of Liberty Bowl. It is a punch bowl made by Revere in 1768 to commemorate the 92 members of the Massachusetts House of Representatives who refused to retract their statement of opposition to the British Townshend Acts of 1767, which imposed taxes on goods imported from England. The punch bowl form is simple but decorated with a considerable amount of engraving including numerous symbols and terms referring to related political events of the day. The defiance of the house members was the first open act of rebellion leading to revolution. This punch bowl has been counted alongside the Declaration of Independence as a significant memorial of the struggle for American independence.

Prior to the outbreak of the Revolutionary War, Revere was also employed as a messenger for the Massachusetts government to carry legislative news to other cities. In 1775 he rode to Lexington to warn John Hancock and Samuel Adams about the British troops that were on their way. The alarm that was raised resulted in the first conflict of the Revolutionary War. In 1861, years after Revere's death, his small part in the event was immortalized in a poem by Henry Wadsworth Longfellow titled "Paul Revere's Ride." During the war, Paul served as a lieutenant colonel in the Massachusetts State Train of Artillery. He was involved in a number of tactical advances prior to leaving the military in 1779.

After the war, the goldsmithing business boomed. People were eager to catch up on the latest fashions and acquire the trappings befitting a new nation. The neoclassical style came into vogue and became synonymous with the democracy of the classical world, which Americans equated with patriotism.

PRESENTATION PIECES

Some of the most interesting and historically significant silver was made for the purpose of commemoration. Paul Revere's famous Sons of Liberty punch bowl is just one example of this wide category of work. Gifts of silver were also presented to war heroes, captains of ships that made long voyages, industrialists, and scholars.

Lieutenant Colonel George Armistead was given a punch bowl in the shape of one of the bombshells that had fallen on Fort McHenry, which he successfully defended in the War of 1812. Upper-class passengers aboard transatlantic ships used to give gifts of silver to the captain after a successful voyage. One or more appreciative students have given silver to a teacher as an act of gratitude. In the eighteenth century, wealthy parents sometimes gave a piece of ceremonial silver to a university and in exchange, any child of theirs in attendance was granted additional privileges. It is from this tradition of presentation silver that some of today's trophies come, including the cup given by Lord Stanley to support a formal hockey championship—the Stanley Cup.

In the Revere shop, much of the silver became standardized with fewer special commissions. This enabled Paul to spend less time supervising the shop, which was generating steady income, and more time in commercial enterprises. Paul started a foundry in 1788 to **cast** cannons, church bells, and ship hardware. He was also the first American to develop a functional copper rolling mill, able to create even sheets of copper for use in metalworking and shipbuilding. This was established in Canton, Massachusetts, around 1800. Such early ventures into industrialization made America less dependent on foreign imports. They were also most likely the reason for the fortune in Revere's possession when he died.

See also Guilds; Nineteenth Century

FURTHER READING

American Revolution, http://www.americanrevolution.org

Forbes, Esther. *Paul Revere and the World He Lived In.* Boston: Houghton Mifflin, 1942.

Goss, Elbridge Henry. *The Life of Colonel Paul Revere.* Boston: Gregg Press, 1972.

Museum of Fine Arts, Boston, and Jonathan L. Fairbanks. *Paul Revere's Boston, 1735–1818* (Exhibition, April 18–October 12, 1975). Boston: Department of American Decorative Arts and Sculpture, Museum of Fine Arts, 1975.

The Paul Revere Memorial Association, Paul Revere House, http://www.paulreverehouse.org

Paul Revere Memorial Association, and Scottish Rite Masonic Museum of Our National Heritage. *Paul Revere: Artisan, Businessman, and Patriot—the Man Behind the Myth.* Boston: Paul Revere Memorial Association, 1988.

Rococo and Neoclassical
(ca. 1700–1800)

In a way, the rococo and neoclassical styles represent different ends of the decorative spectrum. The curving asymmetry and naturalistic references of the rococo contrast with the regular geometry of the neoclassical style that was inspired by decorative motifs from **ancient Greece** and **ancient Rome.** The two styles were both popular in Europe and America during the eighteenth century but enjoyed revivals during the **nineteenth century.** Within those two styles, regional variations also occurred. These were often tied to the political climate of the day. These stylistic changes traveled with artisans such as the Huguenots (see **Paul de Lamerie**), in pattern books and through the objects themselves.

The eighteenth century also saw the change from small workshops to large industrial practices in the silversmith's trade. It was a period of increased demand, weakening **guilds,** more efficient organizational systems, and greater specialization. During the mid-eighteenth century there was an increase in partnerships between metal manufacturers, which amplified divisions between administration and production. This century was also known as the Age of Enlightenment, when new theories regarding the natural world were being tested.

ROCOCO

The rococo followed the **baroque** style in most of Europe. Although both styles incorporated curvaceous elements, the freer and more playful rococo was a lighthearted and intimate response to the bold seriousness and formality of the baroque. The rococo was a style in the fine and decorative arts that initially developed in Paris. The term rococo comes from the French word for artificial shell work, which was *rocaille.* The first examples were seen in the early eighteenth century, and in some areas evidence can be seen into the end of that century. The style spread across Europe and America but was particularly prominent in France, Austria, and Germany.

The characteristics of the rococo style included visual movement, natural forms such as shells and vegetation, curving lines, S- and C-scrolls, asymmetry, and a play of light, which metalwork was particularly well-suited to. Well-known practitioners of the style include silversmiths Paul de Lamerie and Juste-Aurèle Meissonnier.

In the earliest examples of the style, rococo motifs were incorporated through flat chasing on forms that were not necessarily rococo themselves. This was followed by highly dimensional **chased and repoussé** and **cast** elements on more curvaceous forms. Toward its end the rococo style also incorporated piercing, more exuberant foliage, and chinoiserie, an imaginary and Westernized version of **Chinese** subject matter. The rococo metalwork made in England tended to remain restrained.

In silverwork the rococo style meant a lot of work for professional chasers. Metalwork was also incorporated into the furnishings of the day in the form of cast and gilded bronze mounts. Gilded bronze was also used in fireplace accessories, sculptural clocks, and lighting devices. Rococo interiors frequently included much white and gold in the decoration. Its detractors considered the light and ornamental rococo style an expression of courtly excess and unnecessary frivolity.

NEOCLASSICAL

The neoclassical style followed as a reaction to the rococo style. It also initially began in France. Examples can be seen from the mid-eighteenth century forward. At that time European decorative arts were influenced by a resurgence of interest in antiquity. This was partly due to archeological discoveries at Herculaneum and Pompeii. Some experienced Rome firsthand through the Grand Tour, a journey through Europe made by the young and wealthy in order to better their education. The countries undergoing revolutions such as France and America also embraced neoclassicism because of its reference to the classical ideals (see **Paul Revere**). In France it was sometimes called the Gout Grec.

Simpler geometrical forms characterized the neoclassical style with restrained embellishments taken from ancient ceramic and stone work. Classical silver was not discovered until the nineteenth century, so metalworkers in the eighteenth century had to mimic other media. Typical motifs included palmettes, floral swags, wreaths, acanthus leaves, and medallions. These were sparingly used on vessels that tended to have straight lines and frequently incorporated drop ring handles.

As the rococo had been flavored with chinoiserie, so the neoclassical style was influenced by **ancient Egyptian** motifs. In France this was related to the campaign of Napoleon I. As rolling mills became available, the prefabricated flat sheets worked well with straight-sided forms of neoclassical silver vessels. It was during this time that mass production of metal goods had begun in Birmingham and Sheffield. Sheffield Plate (see **Gilding and Plating**) or very thin silver was readily formed into neoclassical metalwork by **stamping and die forming** and by spinning (see **Lathe Work**).

A number of neoclassical substyles developed in several countries. In France the early neoclassical style was also known as Louis XVI or Louis Seize, and the later neoclassical was referred to as Empire style. In England the later neoclassical style was sometimes called Adams, and the later version was called Regency.

LOUIS SEIZE

This style was named for the French monarch who promoted it. It consisted of both a mixture of rococo forms with neoclassical ornamentation and neoclassical forms with rococo decoration. It was more austere than the preceding rococo, with subdued curves and restrained geometry.

EMPIRE

The Empire style lasted for the first quarter of the nineteenth century and was named after the empire that Napoleon Bonaparte had conquered during the first decade. Napoleon favored classical themes in order to symbolically align himself with Roman emperors. The style included references to classical antiquity, images of bees, the letter N, laurel wreaths, eagles, and Egyptian elements referencing Napoleon's Egyptian campaign. As Napoleon conquered more land, the style spread.

ADAMS

The Adams style was named for the influential English architect and designer Robert Adam. His name was synonymous with the neoclassical style as it was expressed in England. Adam had studied in Italy and was particularly influenced by the archeological digs there.

REGENCY

Regency was the name of the English decorative style that began in the last quarter of the eighteenth century and lasted through the first quarter of the nineteenth. The stylistic period was named for the prince regent who was in power at the time and later became George IV. It consisted of heavier Roman forms and incorporated Egyptian motifs without any overt references to Napoleon. Many sculptural presentation pieces and large tablewares with highly detailed decoration were produced in this style. Regency utilized fewer gilded bronze mounts than its French counterpart and was epitomized by the work of silversmith Paul Storr.

During the eighteenth century a number of businesses embraced the increased specialization and mechanization of metalworking to great success. The name Matthew Boulton is synonymous with industrial organization from the period. In addition to more elaborate and traditionally made silver, his company mass produced small metal articles utilizing the newest technical innovations in a pioneering manner. Boulton supported James Watt in his invention of an improved steam engine, which made the first practical steam power. Boulton then harnessed such technical experiments and increased standardization of his wares to great commercial accomplishment.

GOLD AND SILVER

Gold and silver continued their desirability for both their inherent worth and as a status symbol. Much metalwork from the period was remade as styles changed or precious metals were recalled by governments such as France to pay for military campaigns.

By the early eighteenth century, *service à la française,* serving food from the table rather than from a sideboard by servants, was becoming a more common dining practice. Thus the serving platters and tureens became decorative focal points on the table, and much attention was lavished on them. This style of self-service spread across the continent from France, which was the leader in fashion at the time.

BRASS AND BRONZE

The use of brass and bronze hardware for furniture began in the seventeenth century, but reached a high point during the rococo period. These elements began as protective devices. Some acted as bumpers to protect furniture when it was being moved. Escutcheon plates were affixed around keyholes to prevent scratching of the wood with keys. During the rococo these became a part of overall decorative effect of an interior. Additional effects were achieved by gilding, varnishing, and selectively shining the metal. Although there were separate guilds and workshops of furniture makers and brass or bronze mount makers, they collaborated on such furnishings.

IRON

During the rococo the aspiring middle class continued its increased patronage of the blacksmiths in order to procure architectural details such as railings and grills. Such elements were varnished to prevent rust. Iron hardware also continued to be used extensively in the churches (see **Renaissance**).

Quite possibly the most famous decorative ironwork from the rococo period is the ironwork at the Place Stanislas by Jean Lamour. This open space in Nancy, France, includes screens, candelabra, and arches in wrought and gilded iron. From the middle of the eighteenth century, extraordinary rococo ironwork was being produced in Germany and Italy too, continuing the strong traditions from the Renaissance and **Middle Ages.** Following the rococo the use of wrought iron fell off somewhat, as cast iron was becoming easier and cheaper to produce. The versatile cast iron was used in place of many more expensive materials starting in the nineteenth century.

See also Arts and Crafts Movement

FURTHER READING

The Cooper-Hewitt National Design Museum, New York, http://cooperhewitt.org

Council of Europe, Royal Academy of Arts (Great Britain), Victoria and Albert Museum, and Arts Council of Great Britain. *The Age of Neo-Classicism* (catalogue of the Fourteenth Exhibition of the Council of Europe, held at the Royal Academy and the Victoria and Albert Museum, London, September 9–November 19, 1972). London: Arts Council of Great Britain, 1972.

Feld, Elizabeth, and Stuart P. Feld. *Of the Newest Fashion: Masterpieces of American Neo-Classical Decorative Arts: An Exhibition Commemorating the Fiftieth Anniversary of Hirschl & Adler Galleries.* New York: Hirschl & Adler Galleries, 2001.

Grimwade, Arthur. *Rococo Silver, 1727–1765.* London: Faber, 1974.
History Periods and Styles, 18th Century, The Victoria and Albert Museum, London, http://www.vam.ac.uk/collections/periods_styles/18thcentury/index.html
Kimball, Fiske. *The Creation of the Rococo.* Philadelphia, PA: Philadelphia Museum of Art, 1943.
Savage, George. *French Decorative Art, 1638–1793.* New York: Praeger, 1969.

Schwarcz, June
(1918–)

June Schwarcz is a pioneer of contemporary **enameling.** She is best known for sculptural vessel forms created through the use of **electroforming** and combined with glowing transparent enamels. Schwarcz is also an example of the revival of interest in metalsmithing that was part of the post–World War II **Studio Movement.**

June Schwarcz was born June Theresa Morris in Denver, Colorado, in 1918. She studied industrial design at the Pratt Institute in New York and subsequently held a number of jobs in related design fields. In 1943 she married Leroy Schwarcz, whose career as an engineer required several relocations. In 1954, while residing temporarily in Denver, June was able to spend time with an old friend, Terry Touff. Terry was one of several women who were getting together periodically to do some enameling in each other's kitchens as a hobby. Touff and the other women shared the enameling techniques that they knew and introduced June to a book on the subject by Kennth Bates, *Enameling: Principles and Practice.* After leaving Colorado, June continued to work in enamels. Schwarcz and her family eventually settled in California, where she continued to be active in the artistic community and was a founding member of the Northern California Enamel Guild.

As an enamellist, June is largely self-taught. At first, the metal pieces that Schwarcz enameled on were pre-made forms that she purchased. In order to integrate a linear quality in their design, she began to **etch** the surfaces of these forms. Layering transparent enamels over such textured surfaces resulted in a form of *basse taille.* In later forms, she manipulated the readymade metal elements through hammering. June developed an interest in electroforming, which led to her husband's building a 30-gallon electroplating bath for her in their home in 1962 (see **Gilding and Plating**). Rather than use electroplating to create a veneer on metal, June would protect selected areas of a piece and plate excessively on the exposed surfaces. After days of plating, areas that had received the

ALEXANDER FISHER (1864–1936)

Alexander Fisher was a significant figure in the development and promotion of enameling as an art form during the British **Arts and Crafts Movement.** Although the incorporation of enamels was frequently done as a replacement for more expensive gemstones at the time, enamelists such as Fisher were pushing the expressive potential of the medium and treating the enamels more like miniature paintings. His detailed depictions of figures in narrative settings utilized multiple layers of transparent enamels that were fired many times over reflective metal foils. Fisher's most lasting contribution may have been his role as an educator. He lectured extensively and in 1906 published *The Art of Enameling Upon Metal.* A number of his students went on to great critical acclaim as well, including Edith and Nelson Dawson, John Eyre, and Laurin Hovey Martin. Fisher's influence was carried abroad by his students, including Martin, who taught enameling techniques to Kenneth Bates, who has been referred to as the Dean of American Enamelists.

treatment would have built up encrustations of copper with an irregular, organic surface several millimeters higher than the original substrate. The final technical breakthrough in June's work came in 1964 when her husband brought home some thin copper foil for her to experiment with. Finding that the foil could be manipulated by hand much like fabric, June began to fold, gather, and sew the foil. Although the foil alone was not very substantial, the electroplating process allowed for the addition of bulk and substance. This is how June Schwarcz's most original and creative work has been made: through the manipulation of foil followed by selective electroforming and applications of transparent enamels. Her earlier works consisted of bowls and panels. As June's work developed, she continued to make forms that reference the vessel, although they are not considered utilitarian. Their artfulness has placed them beyond the realm of utility. Inspiration for June's work comes from a number of different sources including textiles, modernist painting and sculpture, the folk art of a variety of cultures, and Japanese aesthetics.

See also Verdun, Nicholas of

FURTHER READING

Falino, Jean. "Women Metalsmiths." In *Women Designers in the USA 1900–2000 Diversity and Difference,* ed. Pat Kirkham. New Haven, CT: Yale University Press, 2000.

Jazzar, Bernard, and Harold B. Nelson. *Painting with Fire: Master of Enameling in America, 1930–1980.* Long Beach, CA: Long Beach Museum of Art, 2007.

Schwarcz, June. *June Schwarcz: Forty Years/Forty Pieces.* San Francisco: San Francisco Craft and Folk Art Museum, 1998.

Sentoku. ***See*** Metals

Shakudo ***See*** Metals

Shibuichi. ***See*** Metals

Silver. ***See*** Metals

Silverware. ***See*** Flatware

Sinking. ***See*** Raising

Smelting. ***See*** Mining and Smelting

Soldering. ***See*** Connections

Stamping and Die Forming

Stamping and die forming are both methods of using force to impart a predetermined, dimensional form to metal.

On a small scale, stamping usually involves holding a tool called a stamp, stamping tool, or punch that has a design element on one of the faces. These tools are made of a harder metal than the one being acted on and are generally linear, with the design occupying one of the smaller ends. By holding the decorative end onto the surface of the metal being stamped and striking the opposite end of the stamp with a hammer, the force of the blow causes the stamp to make an impression in the softer metal object. The design is thereby transferred. The repeated use of stamping can create patterns on the surface of an object. One traditional use of stamping is in hallmarking (see **Guilds**). Once the Industrial Revolution provided new forms of mechanical power, hand-stamping processes were adapted to the machine.

A die is a tool that may be used to form or cut metal. A forming die is usually larger than a stamp and is used to form the metal by doing more than making a mere impression. If the design on the die is in relief it is said to be a cameo. If the design is recessed, it is considered an intaglio. A cameo die creates an intaglio design in the metal it acts on, and an intaglio die creates a cameo.

In small-scale production, a piece of metal may be hammered into an intaglio die from the back. Conforming dies are matching cameo and intaglio dies that act on the plane of metal at the same time from both sides to create an effect similar to **chasing and repoussé,** and this process is sometimes referred to as embossing or impressing. When done on an industrial scale, it is a good way to reproduce a dimensional design quickly. However, the making of the dies for embossing is highly skilled work, and the resulting piece is limited in that it cannot contain any undercutting. When this method is used to produce a coin or medallion it is sometimes referred to as coining. This technique has been in use since the days of **Ancient Greece** in order to manufacture coins and similar objects.

The decorative element of a die may be carved or **engraved** into the face or cast in place. Harder metals such as brass, steel, or zinc are used to make tools intended for long-term use. The pressure exerted on dies came initially from hand-held hammers and later from machines such as the drop hammer or fly press. A drop hammer is used to bring greater pressure to bear on the metal by the force of gravity. The hammer or impacting element is raised with a pulley, either by a person or mechanically. When released, gravity acts on it, and the weight of the hammer falls with a force greater than a person could wield with a hand-held hammer. The pressure of the falling element may be also augmented or controlled by mechanical means. A fly press or stamping press exerts pressure in a similar manner, but instead of falling, it twists up and down on a threaded shaft turned by a flywheel. The arrival of steam-powered die forming machines resulted it the manufacture of quickly and inexpensively made **flatware.**

A nonconforming die provides a matrix with a negative space into which metal may be stretched to balloon into the cavity through the use of pressure such as that which may be exerted by a hammer or hydraulic press and urethane. Blanking or cutting dies can be used to cut out a shape from sheet metal. Using a pinching action, two hardened metal edges are made to just pass each other from each side of the metal, similar to the way scissors act. Each side must fit very snugly to work well. The resulting pieces of cutout metal are called blanks.

See also Chasing and Repoussé; Flatware; Forging; Nineteenth Century

FURTHER READING

Kingsley, Susan. *Hydraulic Die Forming for Jewelers and Metalsmiths.* Carmel, CA: 20-Ton Press, 1993.

Untracht, Oppi. *Jewelry Concepts and Technology.* New York: Doubleday, 1982.

Steel. ***See*** Metals

Studio Movement
(ca. 1940–)

During the twentieth century there was a revived interest in manufacture through traditional handwork techniques. Similar to the **Arts and Crafts**

Movement, the Studio Movement (sometimes called the Studio Craft Movement) began partially in reaction to the increased mechanization of its age and the alienation brought by modernity. Its practitioners wanted to be able to control all of the aspects of creation, from design to execution, on a small scale of production. Despite its similarities with the Arts and Crafts Movement, the Studio Movement differed in one very particular way: the studio craftsmen largely considered their work as a means of self-expression. This was as much due to the rising sense of the individual within these societies as to the influence of the fine arts of the period.

After the decline of the Arts and Crafts Movement, handwork in metal was further diminished by the economic downturn of the Great Depression and the absence of raw materials during World Wars I and II. Educational opportunities in metalworking had been reduced to a few technical schools, and there were a limited number of practitioners actively engaged in the making of nonindustrial metalwork.

The Studio Movement had developed in most Western countries by the 1950s. Countries hit hardest by the wars were slower to respond to the new trends. Scandinavia was an early leader in the Studio Movement. Elegant and functional tablewares in silver were made with biomorphic curves that provided more humanism than their **art deco** predecessors had. Sweden in particular emerged as an influence in the new trend, with metalworkers such as Sigurd Persson making one-of-a-kind work in silver as well as designing for the stainless steel industry.

The postwar boom led to a host of new public commissions, especially in Great Britain. The Royal College of Art in London produced craftsmen who were both capable of working for industry and in the niche market of one-of-a-kind, high-end wares. In Germany the emphasis was on technical execution with the apparent continuation of many of the late **Bauhaus** ideals.

In no other country, however, did the postwar Studio Movement take shape as it did in the United States. America had no history of a **guild** system or of a monarchy that commissioned silver plate. The establishment of the country itself came not so very long before the advent of the Industrial Revolution. Therefore, America is a country without a long indigenous metalworking tradition, which may have contributed to a greater sense of freedom and experimentation upon the reintroduction of the practice.

The first stirrings of the Studio Movement in America took place between World Wars I and II. The Works Progress Administration supported many craft practitioners during the Depression by establishing around 3,000 different handicraft projects. During World War II, crafts were learned as hobbies by enlisted men and as therapy by veterans.

Margret Craver was exceptionally important in the development of postwar metalsmithing in the United States. She was interested in learning silversmithing but found that no educational programs existed. She gleaned what knowledge she could from practitioners in the United States and eventually traveled

THE AMERICAN REVIVAL OF BLACKSMITHING

During the early twentieth century, the blacksmith's trade was nearing extinction. Machines had taken many of their jobs, and the use of ornamental wrought iron work was in decline and being replaced by cast iron. The few practicing blacksmiths who remained were usually in rural areas, where they were employed in fixing a variety of farm implements. The rebirth of blacksmithing in America occurred more under the auspices of art than of utility.

Metalsmiths from the Studio Movement, those educated in nonferrous metalwork (see **Metals**) who had been operating since the 1940s, already had college and university programs in place. It was to be through that network of programs that the first academic demonstration of artist-blacksmithing was publicized. Just as many of the modernist jewelers were originally painters, so many of the revivalist blacksmiths were originally jewelers.

L. Brent Kington is frequently hailed as the man responsible for the start of the blacksmithing revival. He was a product of the Studio Craft Movement in jewelry and hollowware, teaching at Southern Illinois University at Carbondale. In 1964 Kington attended the first World Congress of Craftsmen in New York City, sponsored by the American Craft Council. He was struck by the quality of the metalwork in the armor displays at the Metropolitan Museum of Art, and noticed that many of the same techniques that he plied in the nonferrous metals were used in steel. Upon his return to Illinois, he located a blacksmith to teach him the basics of tending a fire and some rudimentary forming. The blacksmiths he initially learned from were in their 70s and 80s, having been agricultural smiths for most of their lives.

In 1970 a book by Alex Bealer, *The Art of Blacksmithing,* came out, and Kington invited Bealer to give a demonstration at his school. Word of the demonstration spread through his connections with the Society of North American Goldsmiths. Fifty people from out of state showed up to see the demo, some from as far away as Colorado and New York. Those who had witnessed the demonstration returned home to build forges of their own.

At a second demonstration, some of the people in attendance decided that what they needed was an organization for those interested in blacksmithing. It was at that time that the Artist-Blacksmith's Association of North America was founded. That event is commonly regarded as the start of the renaissance of blacksmithing in America. Since then a number of outstanding American blacksmiths have risen to international prominence including Albert Paley and Tom Joyce.

to Sweden to study with Baron Erik Fleming. World War II interrupted these lessons, but it also brought a new opportunity. Craver convinced the head of rehabilitation for the armed services hospitals to include silversmithing in their rehabilitation programs.

The greatest impact that Craver had on contemporary metalsmithing was probably the month-long workshops that she persuaded the silver refiners Handy & Harmon to sponsor from 1947 to 1953. These workshops were led by classically trained European silversmiths and were expressly intended to train art educators from colleges and universities in the techniques of silversmithing. The individuals who took those workshops became America's first generation of artist-metalsmiths.

After World War II, many of the returning veterans were disillusioned and wanted a life outside of the traditional office environment. The Servicemen's Readjustment Act of 1944, commonly known as the GI Bill of Rights, provided veterans with help in returning to civilian life. For many this included the opportunity to attend institutions of higher education without restriction on their subjects of study. Numerous returning veterans chose to enroll in art and craft classes, increasing the demand for faculty greatly. Many of the teachers who were hired to fill the shortage knew only a little more than their students. By 1950, college degrees in metalsmithing and jewelry were available in the United States.

The creation and expansion of such programs continued throughout the 1950s and into the 1960s. Thus, generations of makers became craftsmen-teachers who were essentially supported by the state. Consequently, these craftsmen felt an increased freedom to experiment with their media for they did not have to rely on the sale of work to draw a paycheck. Postwar metalsmithing in the United States was a hybrid genre combining the previously disparate practices of goldsmithing, silversmithing, coppersmithing, and eventually blacksmithing.

In the 1960s, the growth of school programs slowed and the availability of teaching jobs decreased. At the time, American society was in the midst of a period of social and political change. This included a back-to-nature counterculture that fit very well with the alternative lifestyles of some studio craftsman. Students continued to enroll in craft programs and upon graduation many sought the more solitary lifestyle of handmade production work for craft fairs. Eventually, more markets opened for metalwork, and some fine art galleries began showing works in craft media. In 1969 the Society of North American Goldsmiths, the primary professional organization of American Studio Metalsmiths, was founded.

In the 1950s the style of metal work being produced in the United States was heavily influenced by the Scandinavian Modern style (see **Georg Jensen**). Sleek, smooth finishes and biomorphic forms prevailed. Objects were designed around traditional, functional formats such as tableware, **flatware,** and liturgical work. In the social and political upheavals of the 1960s, traditions were questioned, and the ties to functional forms that had largely prevailed to this point were shattered by many. The increasingly interconnected field of metalsmiths began to push the limits of the medium as an expressive tool. Although American metalwork owes a great debt of influence to many European and Asian makers, the strong inclination toward greater personal expression and narrative appears to be a largely American trait.

Additional experimentation with new materials and technologies also diversified the practice. Individuals such as **June Schwarcz** began to experiment with **electroforming.** Eugene and Hiroko Sato Pijanowski brought the ancient techniques of *mokumé gane* to the United States from Japan (see **Lamination Techniques**). **Anodized** aluminum started showing up in one-of-a-kind metalwork along with found objects, resins, plastics, and a myriad of other elements from industry. Since the 1980s a large consumer culture has grown up around the metalsmithing field in the form of magazines, galleries, books, and classes.

See also Bauhaus; Colonial America; Modernism

FURTHER READING

The Artist-Blacksmith's Association of North America, http://www.abana.org

Braznell, W. Scott. "The Early Career of Ronald Hayes Pearson and the Post-World War II Revival of American Silversmithing and Jewelrymaking." *Winterthur Portfolio* 34, no. 4 (1999): 185–213.

Fenton, Steve, and Jo Lauria. *Craft in America.* New York, Clarkson/Potter Publishers, 2007.

The National Ornamental Metal Museum, Memphis, Tennessee, http://www.metalmuseum.org

Nordness, Lee. *Objects: USA.* New York: Viking Press, 1970.

The Society of American Silversmiths, http://www.silversmithing.com

The Society of North American Goldsmiths, http://www.snagmetalsmith.org

Stern, Jewel. "Striking the Modern Note in Metal." In *Craft in the Machine Age,* ed. Janet Kardon. New York: Harry N. Abrams, 1995.

Tiffany & Company
(1837–)

The history of Tiffany & Company illustrates the changes that took place in silver manufacture and consumption in the second half the nineteenth century as well as the importance of the international exhibitions and America's arrival as a significant contributor. Although the enterprise began as a stationery goods store in 1837, by the end of that century the name Tiffany was synonymous with style and luxury in metalwork on both sides of the Atlantic.

Charles Lewis Tiffany was the son of a cotton manufacturer. In 1837 at the age of 25 he had accumulated several years of experience running a store for his father in rural Connecticut. Charles had a friend from school named John B. Young who had worked in a stationer's store in New York, and together they decided to try opening their own shop in New York City. They borrowed $500 apiece from Tiffany's father and opened a fancy goods store called Tiffany & Young. Initially the store sold a variety of imported decorative items, accessories, and stationery, including papier-mâché, dressing cases, umbrellas, small furniture, and pottery.

In 1841 the firm became Tiffany Young & Ellis when J. L. Ellis was added to the partnership. Although the company had retailed a few silver articles, it was not until 1848 that they began to engage in the manufacture of silver wares. It was in that year that they hired Gustav Herter as a silver designer. He worked for the company for three years. Following this, Tiffany Young & Ellis made an exclusive arrangement to sell the hollowware of popular local silversmith John C. Moore. The business would expand to include flatware in 1869. Although there was no American standard of silver at the time, the company produced the equivalent of British sterling, a higher quality silver alloy than many American manufacturers were producing at the time. This later became the standard silver alloy in the country (see **Metals**). In 1853 the partners Ellis and Young retired from the company and the name was changed to Tiffany & Company, which it remains to this day. Later, electroplated wares were produced as well, in a plant located in New Jersey (see **Gilding and Plating**).

Soon after the agreement was established with John C. Moore, he retired, and his son, Edward C. Moore, took over. This was to be the beginning of Tiffany & Company's ascension to the world stage. The popularity of Moore's silver wares caused greater demand and expansion of the work and retail spaces. The growth of the silver shop also increased the division of labor and specialization of each worker in a manner similar to that which was taking place across many industries at the time. At its height the workshop employed some 500 men.

The firm's participation in international expositions helped to increase their public recognition. The Paris Exhibition of 1867 was the first one the company participated in. They showed their standard wares, and Moore's designs received a merit award, the first such prize given to an American silver company in an international exhibition.

Moore began to look at and collect Near and Far Eastern art in the 1860s. The influence of those aesthetics started to appear in his designs later in that decade. Initially the silver forms were Western with some superficial Eastern ornament, such as chasing. Applied ornaments in different metal alloys followed this, and finally this culminated in the adoption of Far Eastern forms, usually derived from ceramics. It wasn't until these new influences were developed that Tiffany began to differ significantly from their competitors both at home and abroad.

At the 1876 Philadelphia Centennial Exhibition, Tiffany's contributions displayed Near Eastern inspiration, inlaid with **niello** and copper. The style, called Saracenic, peaked at the 1889 Exhibition in Paris where Tiffany & Company received the grand prize for both those wares and their Japanesque pieces. Many of the Saracenic objects at this exhibition included detailed **enamels** with a satin finish.

The **Japanese** influence can be seen in the silver of Tiffany & Company in the last three decades of the nineteenth century. It was revolutionary in that it was not a historicizing style and in the way its simplicity prefigured the reduced aesthetics of **modernism.** What was also important was the move away from European stylistic leadership. The Japanesque wares that Tiffany & Company displayed at the Paris International Exhibition of 1878 included *mokumé gane* (see **Lamination Techniques**). Several European museums as well as the Japanese government itself purchased these Japanesque objects. Tiffany & Company was awarded the grand prize for silverware, and the company was given appointments to some 23 crowned heads in Europe.

THE POLITICS OF SILVER

Precious metals have a long history of being political instruments. As bearers of meaning they are often employed for reasons of bureaucratic significance in addition to their inherent preciousness. **Paul Revere**'s Liberty Bowl is an excellent example of the relationship between metals and politics during the American Revolution. When the Civil War raged in America from 1861 to 1865, Charles Lewis Tiffany advertised military equipment for sale to loyalist states only. It was also Tiffany & Company who made the much-publicized Gladstone Testimonial. This was purchased with funds raised by Joseph Pulitzer to commemorate English Prime Minister William Gladstone's efforts to gain home rule for Ireland.

One of Tiffany & Company's most popular **flatware** patterns called "Japanese" was renamed "Audubon" during World War II, after John James Audubon, who was known for his studies of North American birds. The new reference was more in keeping with the political sympathies of the day.

Of course Moore was not responsible for all the objects that came from Tiffany & Company. The company maintained a training program for designers and a design library. Other styles that came out of Tiffany & Company around the turn of the century included Renaissance Revival, Native American, **Celtic,** and such strange concoctions as Burmese, Pompeian, and Viking. This was in keeping with the eclecticism of the **nineteenth century.**

Some well-known designers that trained under Edward C. Moore included John T. Curran and Paulding Farnham. Other important Tiffany designers included Charles T. Grosjean and Eugene Soligny. Tiffany also took advantage of the number of skilled immigrants, trained in European manufactories, who came to America. James Whitehouse of England was one such person. He had been an

engraver and designer in his native England and ended up designing the revised face of the Great Seal of the United States at Tiffany & Company.

Following the Civil War, enormous amounts of silver were being discovered in Nevada and California, leading to a drastic reduction in price. The late nineteenth century was a time when many fortunes were made. One of the acceptable ways of showing off one's wealth was in silver for use and display. The silver boom, the rise of the newly moneyed class, and Tiffany & Company all met in the creation of the famous Mackay silver service. John Mackay, called the "silver king," had started as a poor Irish immigrant and was propelled into wealth and fame when his mine became one of the greatest silver and gold producers in Nevada, nicknamed the Big Bonanza. It has been said that the half-ton of silver that went into the set came directly from the mine to Tiffany & Company in New York. The service included 1,250 pieces and was meant to serve all the household silver needs, from intimate teas to elaborate dinner parties for 24. The service was completed in 1878 and had taken 200 silversmiths two years to make. The design incorporated Saracenic and naturalistic motifs and was displayed at the 1878 Paris Exposition.

Tiffany & Company has also been responsible for an array of famous presentation pieces (see **Colonial America**). These include the Bryant Vase and the Adams Vase as well as renowned sporting trophies such as the Vince Lombardi Super Bowl trophy, the World Series Cup, and the trophy of the National Basketball Association Championship.

In the twentieth century the demand for silver declined as the structure of the home changed and the price of silver increased. Tiffany & Company continued to employ important designers such as Frank Lloyd Wright and Elsa Peretti. The company has a significantly diminished part in the manufacture of silver today but continues to be known for its high level of craftsmanship and splendid objects.

See also Islamic Metalwork; Rococo and Neoclassical; Yellin, Samuel

FURTHER READING

Blades, John M., and John Loring. *Tiffany at the World's Columbian Exposition.* Palm Peach, FL: Henry Morrison Flagler Museum, 2006.

Carpenter, Charles Hope, and Janet Zapata. *The Silver of Tiffany & Co., 1850–1987.* Boston: Museum of Fine Arts, 1987.

Carpenter, Charles Hope, and Mary Grace Carpenter. *Tiffany Silver.* New York: Dodd, Mead, 1978.

Loring, John. *Magnificent Tiffany Silver.* New York: Harry N. Abrams, 2001.

The Metropolitan Museum of Art, New York City, http://www.metmuseum.org

Tiffany & Company, http://www.tiffany.com

Tin. ***See*** Metals

Titanium. ***See*** Metals

Verdun, Nicholas of
(active 1181–1205)

Nicholas of Verdun was one of the most significant craftsmen in medieval Europe. A French goldsmith and **enamelist,** his work embodies the transitional style that occurred in the arts between the Romanesque and Gothic periods (see **Middle Ages**). Two signed works of his have survived, with numerous others attributed to him.

The church and its monasteries were commissioning most of the artwork made at the end of the twelfth century. In metalwork, enamels were used a great deal as a means of enriching surfaces. Small enameled objects were executed in gold. Larger pieces were made in bronze or copper that was then **gilded.** The major enameling centers at the time were in Spain, the Rhine River area, the Meuse River area, and Limoges, France. Created in monasteries and workshops, the enameled pieces were meant to serve specific religious functions.

Because there are no written records of his life, all we know of Nicholas is what we can determine from the metalwork he left behind. His style has led scholars to believe that he was trained in the Mosan school of metalworking in the Rhineland area. It is considered the first great school of enamelists in Western Europe as well as a regional pioneer of the champlevé technique. Nicholas was an itinerant craftsman, moving to wherever patrons employed him. Because of the significance of the commissions he was given, we may assume that he had built up a significant reputation for himself prior to undertaking his two known, signed works.

The earliest known work by Nicholas of Verdun is an altarpiece covered in enameled plaques in the Klosternueberg Abbey near Vienna. It was completed in 1181 and signed "Nicolaus Virdunensis." Originally, these plaques were on an ambo, a pulpit from which parts of the service were read. The original 45 champlevé enamel plaques were remade into a three-part altarpiece after a fire in 1330. Six additional plaques were added in order to accommodate the new form. It is the largest and most important surviving enamelwork from the Middle Ages.

The metal of the champlevé plaques is gilded copper. Each plaque contains a single scene from the Bible depicted within a trefoil-arched space. The enameled background is dark blue with linear details in red, turquoise, and white. The scenes are arranged in the typological manner, which illuminates the relationships between different biblical events. The imagery is arranged in three rows: the top includes scenes from the Old Testament, those in the central row are from the life of Jesus, and the bottom row is made up of scenes from the New Testament. Vertically, the specific episodes are arranged according to common themes and theological symbolism. The only deviation in this organization occurs at the end in the Last Judgment. Each panel also bears an inscription, and between the arches are representations of angels, embodiments of the virtues, and figures of the prophets. Taken as a whole, the altarpiece is a sermon embodied in gold and glass. Although this method of ordering the subject matter according to theological symbolism was being incorporated in many art forms at the time, the Klosternueberg altarpiece is the most extensive use of such a typological program to have survived from the period.

There is also a visible stylistic progression in the piece. Moving from left to right one can see gradual changes in the treatment of the figures and their drapery. The first plaques made have the characteristics of the Romanesque period. They are restrained, delicate in style, and static. Moving to the right, the plaques show progressively looser and more naturalistic figures with increasingly expressive postures and a bolder line quality. A project of this scope would have taken many years to complete, during which time a level of naturalism that is unusual for the time becomes evident. This new approach is sometimes referred to as the transitional style: a style that began to develop at the same time as the Gothic but with a greater emphasis on naturalism and classicized human forms.

The second signed piece by Nicholas of Verdun is the reliquary of Saints Piatus and Nicasius in the Cathedral of Tournai in Belgium, signed "Magister Nicolaus de Verdun." It is sometimes referred to as the Shrine of the Virgin Mary, and was completed in 1205. Reliquaries are ornate containers created to house the remains or related artifacts of famous religious figures, such as saints. This example is a chasse, or reliquary in the form of an architectural structure. It was made with gilded silver and bronze with enamel and gemstones. In this piece, dimensional chasing and repoussé create the main imagery, and flat, enameled areas simply serve as background. Although the reliquary is more Gothic in style than the Klosternueberg altarpiece, it is still transitional in its use of naturalistic figures.

See also Middle Ages

FURTHER READING

Lasko, Peter. *Ars Sacra, 800–1200.* New Haven, CT: Yale University Press, 1994.
Ross, Leslie. *Artists of the Middle Ages.* Westport, CT: Greenwood Press, 2003.
Wessel, Klaus. *Byzantine Enamels, from the 5th to the 13th Century,* Irene R. Gibbons, trans. Greenwich, CT: New York Graphic Society, 1968.

White Metal. ***See*** Metals

Yellin, Samuel
(1885–1940)

Samuel Yellin was arguably the greatest blacksmith in the history of the United States. Classically trained in Europe, this immigrant was a major part of the **Arts and Crafts Movement** in America as well as the revival movements in architecture. Evidence of his work can be seen in major institutions throughout the United States. His thorough understanding of the practice and history of his craft is unsurpassed.

Samuel Yellin was born in 1885 in Galicia, Poland. His early education exposed him to several different areas of the mechanical arts, and at the age of 11, a Russian blacksmith took Samuel as an apprentice. He completed the apprenticeship when he was 17 and departed to tour Europe for several years. This journeyman period was a common element of an apprentice's training and enabled young men to learn more about the trade from other masters (see **Guilds**). It is believed that Yellin traveled in Germany, Belgium, France, and England, all the while earning his keep through his craft. In 1906 he immigrated to Philadelphia in the United States, where his mother and sisters already lived. At the time, he was unable to read or write English. In his first three yeas in the United States, Samuel held a variety of jobs in the metal industries while saving money for his own shop.

The Arts and Crafts Movement in America was growing in popularity and led to Yellin teaching classes in ironworking at the Pennsylvania Museum and School of Industrial Art in Philadelphia (now the Philadelphia College of Art). He was an instructor there until 1919. Teaching helped to improve Yellin's communication skills, as did marrying Leah Josephs in 1913.

In 1909, with $150 saved, Samuel Yellin was finally able to open his first shop. It consisted of two rooms on the fourth floor of a building with low ceilings. After three years, he and two assistants moved to a larger space on Jefferson Street. It was in 1915 that he settled at a permanent location on Arch Street. With growing success and recognition, Yellin was able to commission architects from the firm of Mellor, Meigs, and Howe to design a new building for his workshop. This was the space in which all of Yellin's major commissions would be born.

Along with the workshops, there were two special rooms dedicated to the art of metalworking at Arch Street. One was a library: a vast store of research on and images of metalwork. The other room was a kind of museum of ironwork. In addition to samples of wrought iron by Yellin himself, it included historic examples of the craft from all over Europe. These books and objects served as inspiration to

Yellin and his men. Throughout his career, this interest in the history of his field as well as his thorough understanding of the continuum from which his work came gave Yellin an enlightened perspective. This was something he imparted to his students as well as the people who commissioned him. Museums would seek his council for their collections, and he was even asked to contribute as an authority to the *Encyclopedia Britannica* on the subject of ironwork.

During the boom of construction following World War I, Yellin's shop created a great variety of wrought ironwork. Public architecture at the time was a mix of revival styles, which Yellin was eminently capable of designing for by taking into consideration each style's historic precedent as well as the more modern needs. From the English Gothic style to the French Revival, Yellin was versed in it all. His shop produced countless gates, architectural elements, fireplace accoutrements, and light fixtures among other things. His clientele included prestigious academic institutions, churches, and the homes of the wealthy. Harvard University, Washington Cathedral, and the Henry Clay Frick House (now the Frick Museum) all contain work from Yellin's shop. Somehow, no matter how many commissions were undertaken, Samuel Yellin was involved at every stage of their design and execution. Many projects started with a "sketch" executed in iron by Yellin himself.

Yellin's greatest commission was for the Italian Revival ironwork of the Federal Reserve Bank in New York, from 1919 to 1924. His workforce grew to 200 men in order to manufacture the 200 tons of wrought ironwork ordered by the architects. Fellow immigrants who had also been trained in Europe made up most of the employees running some 60 forges. It remains the largest job of its kind in the United States.

The Great Depression led to a decline in the building trades, furthered by the onset of World War II. Although the Yellin workshop did not close, it significantly diminished in size. When Yellin passed away in 1940 his wife and son continued the business. Even today, Samuel Yellin Metalworkers is still in existence in Pennsylvania under the direction of his granddaughter, Clare.

See also Colonial America; Forging; Middle Ages; Nineteenth Century; Studio Movement

FURTHER READING

Andrews, Jack. *Samuel Yellin, Metalworker.* Ocean Pines, MD: Skipjack Press, 2000.

Davis, Myra Tolmach. *Sketches in Iron: Samuel Yellin, American Master of Wrought Iron, 1885–1940.* Washington, DC: George Washington University, 1971.

Fletcher, Tom. "Tom Fletcher's New York Architecture, Images, and Notes," http://www.nyc-architecture.com/ARCH/ARCH-Yellin.htm

Kardon, Janet, ed. *Revivals! Diverse Traditions.* New York: Harry N. Abrams, 1996.

Samuel Yellin Metalworkers, http://www.samuelyellin.com

Wattenmaker, Richard. *Samuel Yellin in Context.* Flint, MI: Flint Institute of Arts, 1985.

Selected Bibliography

GENERAL WORKS

Aitchison, Leslie. *A History of Metals.* New York: Interscience Publishers, 1960.
Benjamin, Susan. *Enamels: The Smithsonian Illustrated Library of Antiques.* New York: Cooper-Hewitt Museum, 1983.
Fleming, John, and Hugh Honour. *The Penguin Dictionary of Decorative Arts.* London: Viking, 1989.
Haedeke, Hanns-Ulrich. *Metalwork,* Vivienne Menkes, trans. New York: Universe Books, 1970.
Knauth, Percy. *The Metalsmiths.* New York: Time-Life Books, 1974.
Smith, Cyril Stanley. *A History of Metallography: The Development of Ideas on the Structure of Metals Before 1890.* Cambridge, MA: MIT Press, 1988.

MUSEUM COLLECTIONS

African Art Museum of Maryland, http://www.africanartmuseum.org
Armory of the State of Styria, Austria, http://www.zeughaus.at
Art Museum at The Chinese University of Hong Kong, China, http://www.cuhk.edu.hk/ics/amm
Asia Society Museum, New York, http://www.asiasocietymuseum.com
The Bauhaus Archives, Berlin, http://www.bauhaus.de/english/
The Bauhaus Foundation Dessau, http://www.bauhaus-dessau.de/en/index.asp
The British Museum, London, http://www.britishmuseum.org
Bröhan Museum, Berlin, Germany, http://www.broehan-museum.de
Colonial Williamsburg, Williamsburg, Virginia (includes The DeWitt Wallace Decorative Arts Museum), http://www.history.org
The Danish Museum of Art and Design, http://kunstindustrimuseet.dk/en
Deutsches Museum, Munich, Germany, http://www.deutsches-museum.de
Dumbarton Oaks, Washington, DC, http://www.doaks.org
Freer Gallery of Art and Arthur M. Sackler Gallery, Smithsonian Institution, Washington, DC, http://www.asia.si.edu
The Hermitage Museum, St. Petersburg, Russia, http://www.hermitagemuseum.org

Hong Kong Museum of Art, Hong Kong, China, http://www.lcsd.gov.hk/CE/Museum/Arts/english/intro/eintro.html
Hopewell Furnace, National Historic Site, Pennsylvania, http://www.nps.gov/hofu
Jewish Museum, Berlin, Germany, http://www.juedisches-museum-berlin.de/site/EN/homepage.php
Jewish Museum, London, UK, http://www.jewishmuseum.org.uk
The Jewish Museum, New York, http://www.jewishmuseum.org
Kunsthistorisches Museum, Vienna, Austria, http://www.khm.at/homeE/homeE.html
Los Angeles County Museum of Art, http://www.lacma.org
The Metropolitan Museum of Art, New York, http://www.metmuseum.org
Museum of Gold, Bogotá Colombia, http://www.banrep.gov.co/museo/eng/home4.htm
National Gallery of Art, Washington, DC, http://www.nga.gov
National Museum, New Delhi, India, http://www.nationalmuseumindia.gov.in
National Museum of African Art, Smithsonian Institution, http://www.nmafa.si.edu
National Museum of Anthropology, Mexico City, Mexico, http://www.mna.inah.gob.mx
The National Ornamental Metal Museum, Memphis, TN, http://www.metalmuseum.org
Nihon Mingeikan (Japanese Folk Craft Museum), http://www.mingeikan.or.jp
Palace Museum, Beijing, China, http://www.dpm.org.cn/english/default.asp
The Palace of Versailles, Versailles, France, http://www.chateauversailles.fr/en/index.php
Peabody Museum of Archaeology and Ethnology, Harvard University, Cambridge, Massachusetts, http://www.peabody.harvard.edu
Petrie Museum of Egyptian Archaeology, University College London, http://www.petrie.ucl.ac.uk
Philadelphia Museum of Art, http://www.philamuseum.org
The Renwick Gallery of the Smithsonian American Art Museum, http://americanart.si.edu/renwick/index.cfm
Royal Armories, Great Britain, http://www.royalarmouries.org
Shanghai Museum, Shanghai, China, http://www.shanghaimuseum.net
Shelburne Museum, Shelburne, Vermont, http://www.shelburnemuseum.org
Victoria and Albert Museum, London, http://www.vam.ac.uk
Winterthur Museum and Country Estate, Winterthur, Delaware, http://www.winterthur.org

PROFESSIONAL ASSOCIATIONS

The Artist-Blacksmith's Association of North America, http://www.abana.org
The Enamelist Society, http://enamelistsociety.org
The Goldsmiths Company, London, England, http://www.thegoldsmiths.co.uk
The Society of American Silversmiths, http://www.silversmithing.com
The Society of North American Goldsmiths, http://www.snagmetalsmith.org

TECHNICAL REFERENCES

Ammen, C. W. *Metalcasting.* New York: McGraw-Hill, 2000.
Andrews, Jack. *New Edge of the Anvil: A Resource Book for the Blacksmith.* Drexel Hill, Pennsylvania: SkipJack Press, 1994.
Briggs, Charles W., Ivan H. Griffin, and Edward M. Roden. *Basic Welding Techniques.* New York: Litton Educational Publishing, 1977.
Curtis, Leslie. *Electroforming.* London: A & C Black, 2004.
Darty, Linda. *The Art of Enameling.* New York: Lark Books, 2004.

Finegold, Rupert, and William Seitz. *Silversmithing.* Radnor, PA: Chilton Book Co., 1983.
Hughes, Richard. *The Colouring, Bronzing and Patination of Metals.* New York: Watson-Guptill Publications, 1991.
Lewis, Marcia. *Chasing: Ancient Metalworking Technique with Modern Applications,* second edition. Long Beach, CA: LarMar Productions, 2000.
McCreight, Tim. *The Complete Metalsmith,* revised edition. Worcester, MA: Davis Publications, 1991.
Midgett, Steve. *Mokume Gane: A Comprehensive Study.* Franklin, NC: Earthshine Press, 2000.
Untracht, Oppi. *Jewelry Concepts and Technology.* New York: Doubleday, 1982.
Untracht, Oppi. *Metal Techniques for Craftsmen,* first ed. Garden City, NY: Doubleday, 1968.

Index

Page numbers in **boldface** type refer to main entries.

About the Author

ANA M. LOPEZ is a metalsmith and educator living in Fort Worth, Texas. She holds both an MFA in metalsmithing from Cranbrook Academy of Art in Bloomfield Hills, Michigan, and a master's degree in the history of American decorative arts from the Parsons School of Design and The Smithsonian Associates in Washington, DC. She is currently teaching undergraduate and graduate courses in metalsmithing and jewelry as well as the history of craft at the University of North Texas in Denton.